Advances in Experimental Medicine and Biology

Volume 1493

Advances in Experimental Medicine and Biology provides a platform for scientific contributions in the main disciplines of the biomedicine and the life sciences. This series publishes thematic volumes on contemporary research in the areas of microbiology, immunology, neurosciences, biochemistry, biomedical engineering, genetics, physiology, and cancer research. Covering emerging topics and techniques in basic and clinical science, it brings together clinicians and researchers from various fields.

Advances in Experimental Medicine and Biology has been publishing exceptional works in the field for over 40 years, and is indexed in SCOPUS, Medline (PubMed), EMBASE, Reaxys, EMBiology, the Chemical Abstracts Service (CAS), and Pathway Studio.

2023 CiteScore: 5.9

Mohammed S. Razzaque
Editor

Vitamin D Function

From Mineral Ion Regulation to Infection

Springer

Editor
Mohammed S. Razzaque
University of Texas Rio Grande Valley (UTRGV)
Edinburg, TX, USA

ISSN 0065-2598 ISSN 2214-8019 (electronic)
Advances in Experimental Medicine and Biology
ISBN 978-3-032-04356-6 ISBN 978-3-032-04357-3 (eBook)
https://doi.org/10.1007/978-3-032-04357-3

This Springer imprint is published by the registered company Springer Nature Switzerland AG
The registered company address is: Gewerbestrasse 11, 6330 Cham, Switzerland

Preface

I am pleased to present a book entitled *Vitamin D Function: From Mineral Ion Regulation to Infection*. Vitamin D is a pleiotropic hormone that plays a critical role in a wide array of skeletal and extraskeletal functions in the human body. In addition to its well-established role in bone formation, it is also involved in the regulation of muscle functions and neuronal activities. The vitamin D receptor (VDR) is expressed in nearly all human tissues, underscoring the importance of maintaining optimal vitamin D levels for essential cellular functions. Dysregulation of vitamin D signaling has been implicated in various pathological conditions, including autoimmune diseases, infections, and tumorigenesis.

This book features contributions from authors who have explained various aspects of vitamin D function and mineral ion regulation in both health and disease. The illustrated chapters provide a comprehensive overview of current knowledge, covering topics from the physiological regulation of vitamin D homeostasis to the consequences of vitamin D deficiency and toxicity and offering insights into future clinical perspectives. Based on the thematic focus, the chapters in this book can be broadly categorized into three areas: (1) detailing diseases associated with vitamin D deficiency, (2) explaining the roles of magnesium and calcium in interorgan communication, and (3) discussing the potential of vitamin D-induced antimicrobial peptides (AMPs) in combating viral and bacterial infections.

Although significant progress has been made in understanding the synthesis and activation of vitamin D, there remains a lack of consensus regarding the long-term health benefits of vitamin D supplementation in humans. I hope that this book will promote greater interdisciplinary collaboration between basic and clinical scientists to advance our understanding of the clinical utility of vitamin D supplementation in otherwise healthy individuals. The time and effort I have invested in compiling this book will be worthwhile if it inspires young physicians and scientists to pursue innovative approaches to understanding not only the roles and regulatory mechanisms of vitamin D and mineral ions in human physiology but also their true therapeutic value as supplements.

I would like to take this opportunity to sincerely thank each of the contributors to this book for generously sharing their knowledge and expertise. Finally, I want to express my heartfelt appreciation to my family (Rafi, Yuki, Lisa, Newaz, Zahid, Shahed, Muhit, and my mother, Nilufar Begum) for their support and encouragement, which helped me bring this book project to completion.

Edinburg, TX, USA Mohammed S. Razzaque

Contents

Vitamin D Activation and Mineral Ion Regulation

Mohammed S. Razzaque

Abstract

Nutrients frequently act in coordination with the absorption and metabolism of one nutrient often interdependent on the presence of others. For example, magnesium is essential for the activation of vitamin D, which, in turn regulates calcium and phosphate homeostasis, both of which are critical for cellular and organ function. A comprehensive understanding of disorders arising from disruptions in vitamin D and mineral ion metabolism is essential for explaining the pathophysiological mechanisms associated with their imbalance. The chapters of this book are selected to provide in-depth insights and highlight important areas of bidirectional relationship between vitamin D and mineral ions for future research to advance the understanding of the delicate interactions between nutrients and overall health.

Keywords

Vitamin D · Magnesium · Calcium · Fracture healing · Sleep disorders · Autoimmune disorders · Antimicrobial peptides

M. S. Razzaque (✉)
Department of Medical Education, School of Medicine, University of Texas Rio Grande Valley (UTRGV), Edinburg, TX, USA

University Kuala Lumpur Royal College of Medicine Perak (UniKL RCMP), Ipoh, Perak, Malaysia
e-mail: mohammed.razzaque@utrgv.edu

This book is intended to combine the roles and regulations of vitamin D and mineral ions in various human diseases. The book includes a total of 11 chapters (excluding this introductory chapter), each of which explores different aspects of vitamin D and mineral ions in both health and disease conditions. Nine of those chapters detail different functional aspects of vitamin D, while two other chapters explain the delicate role of magnesium in maternal-fetal growth and the importance of calcium in the gut-glucose axis. Vitamins and micronutrients, including mineral ions and trace elements, are vital for maintaining the normal physiological functions of various systems and organs [1]. Vitamin D activation and mineral ion regulation are interconnected physiological processes essential for maintaining calcium and phosphate homeostasis in humans [2], and these nutrients play key roles in musculoskeletal and neuronal functions, as well as vital cellular activities and signaling. The various steps of vitamin D synthesis are detailed in a few chapters, with the importance of magnesium in activating its synthesis in the liver and kidneys. All the enzymes that metabolize vitamin D require magnesium, which acts as a cofactor in enzymatic reactions to generate bioactive vitamin D [3–5] (Fig. 1).

Vitamin D is a pleiotropic hormone that has both skeletal and extraskeletal functions. Vitamin D facilitates the intestinal absorption of calcium and phosphate, supports the mineralization of osteoid tissue in bone, and helps maintain serum

M. S. Razzaque (ed.), *Vitamin D Function*, Advances in Experimental Medicine and Biology 1493,
https://doi.org/10.1007/978-3-032-04357-3_1

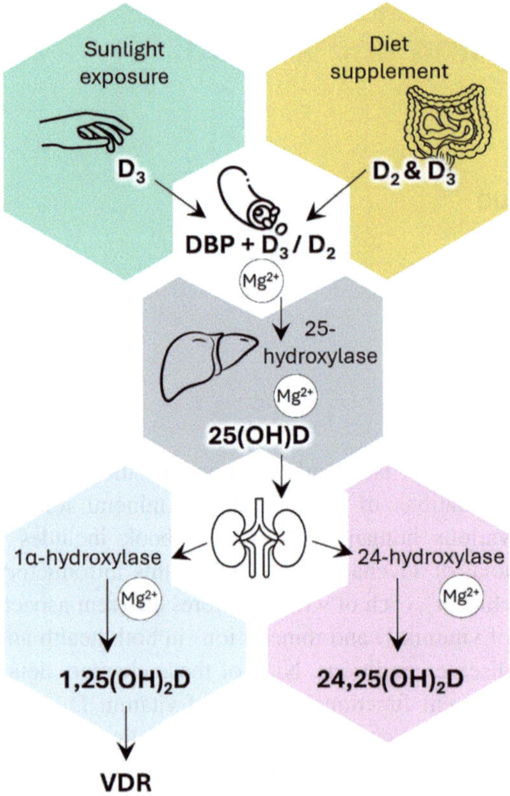

Fig. 1 Role of magnesium in vitamin D synthesis. Abbreviations: **D3**, vitamin D3 (from animal sources); **D2**, vitamin D2 (nonanimal sources); **DBP**, vitamin D-binding protein; **25(OH)D**, calcifediol (inactive form); **1,25(OH)2D**, 1,25-dihydroxyvitamin D (biologically active form); **24,25(OH)2D**, 24,25-dihydroxyvitamin D; **Mg**, magnesium; **VDR**, vitamin D receptor

ionized calcium and phosphate levels. Zaman and Ullah described the role of vitamin D in supporting bone growth and fracture healing, possibly by enhancing immunoinflammatory responses and the mineralization process [6]. Vitamin D exerts anabolic effects on fracture healing by promoting bone mineralization and modulating the activity of bone-forming and bone-resorbing cells to ensure effective fracture repair and bone regeneration [7, 8]. In a separate chapter, the importance of maintaining vitamin D homeostasis in sleep disorders, with clinical evidence, is provided to improve the quality of sleep and overall health and well-being [9]. Vitamin D is likely to influence sleep quality through several biological mechanisms, including regulating the sleep-wake cycle (circadian rhythm), facilitating the synthesis of melatonin, exerting anti-inflammatory effects, and affecting mood and emotional behaviors, although research is still ongoing to identify the exact underlying mechanisms [10, 11]. In a chapter devoted to the role of vitamin D in autoimmune disorders, Fontes-Dantas et al. elegantly detailed the complex relationships among the levels of serum vitamin D, genetic variations, and immune regulation in various autoimmune disorders [12]; the authors emphasized the importance of a deeper understanding of these interactions to develop an innovative strategy to prevent or delay the occurrence of autoimmune conditions. Moreover, vitamin D is likely to play a regulatory role in the immune system, and its inadequacy may, in turn, affect the evolution, progression, or severity of autoimmune disorders [13]. Ahmed et al. detailed the potential neuroprotective effects of vitamin D, possibly by reducing pro-inflammatory cytokines, promoting mitochondrial health, and lowering oxidative stress [14]. In a separate chapter, Mumena et al. explained the prognostic significance of serum vitamin D levels in oral cancer [15]; in a well-illustrated chapter, the authors discussed how the levels of various vitamin D components, including vitamin D receptor (VDR), CYP27B1 and CYP24A1, may be useful indicators of the biological activity of tumor cells and hence predict tumor growth and progression. Vitamin D may influence oral tumor development and progression by exerting antiproliferative and proapoptotic effects and anti-inflammatory and immune-modulating functions and possibly by inhibiting angiogenesis and metastasis; however, more clinical research is needed to determine whether vitamin D supplementation has therapeutic value in delaying oral tumorigenesis [16, 17]. A chapter is also included where the roles and regulations of vitamin D have been described in the context of mineral ion metabolism and potential regulatory molecules [18]. Importantly, vitamin D can induce the skeletal synthesis of fibroblast growth factor 23 (FGF23) to lower phosphate levels in the blood by reducing renal phosphate reabsorption and increasing urinary excretion. Vitamin D can also induce klotho, a coreceptor for FGF23, to facilitate its binding with fibroblast growth factor receptors (FGFR) to control phosphate homeostasis [2, 19, 20]. A regulatory feedback loop exists between FGF23 and vitamin D metabolism. Specifically, 1,25-dihydroxyvitamin D_3 [1,25$(OH)_2D_3$] stimulates the synthesis and secretion of FGF23 by bone tissue, whereas FGF23, in turn, suppresses the production of 1,25$(OH)_2D_3$ by inhibiting the expression of 1α-hydroxylase (CYP27B1) [2].

Lanehart et al. explained why maintaining magnesium balance, a cofactor for vitamin D synthesis, during pregnancy is crucial for both maternal and fetal health [21]; the authors listed the potential complications of magnesium inadequacy for both mothers and infants, particularly in low-income regions, and recommended how public health programs focused on nutritional maneuvers could help reduce complications and improve health outcomes for both mothers and infants. In a separate chapter by Foamkom et al., the potential role of another important mineral ion, calcium, in the gut-glucose axis was briefly discussed [22]. Akimbekov et al. elegantly illustrated how vitamin D regulates the biosynthesis and activity of antimicrobial peptides (AMPs), particularly cathelicidin and defensins [23]; the authors emphasized the interplay between vitamin D and AMPs, highlighting recent advances in understanding the synergistic protective effects and potential clinical benefits in combating viral and bacterial infections [23, 24]. Finally, a chapter by Vasil and Razzaque has discussed the significance of new vitamin D guidelines that have undergone a significant shift from more aggressive vitamin D intake and broader testing recommendations in 2011 to a more conservative, targeted, and evidence-based approach in 2024 [25]; this chapter has also discussed why such a shift might be necessary to minimize the overuse or abuse of vitamin D supplements. Although adequate vitamin D is necessary, more is not always better, as there is a possibility of complications of hypervitaminosis D without elevated serum 25(OH)D levels in individuals with unrestricted consumption of vitamin D supplements [26].

The chapters in this book highlight the significant clinical and biological roles of vitamin D, along with various minerals, including calcium, phosphate, and magnesium, in maintaining homeostatic balance and exerting protective effects that may help prevent or delay the onset of the wide range of disorders listed above. As detailed in the accompanying chapters, adequate mineral ion and vitamin D balance are essential for maintaining overall health. A deeper understanding of disorders linked to disruptions in vitamin D and mineral ion metabolism is needed to better comprehend the conditions associated with their imbalance [27–31]. Furthermore, identifying populations at risk for nutrient inadequacies and offering better guidance on consuming diets rich in vitamins and minerals, along with lifestyle modifications, could delay the onset of various disorders and open new avenues for preventive medicine. I hope this book offers valuable insights and highlights areas that require further research to better understand the complex relationships between nutrients and overall health and well-being. The time and effort I have invested in compiling this book will be worthwhile if it inspires young physicians and scientists to pursue innovative approaches to understand not only the roles and regulatory aspects of vitamin D and mineral ions in human physiology but also the true therapeutic value of vitamin D supplementation in otherwise healthy individuals to avoid potential overuse or misuse.

Acknowledgments I want to thank Dr. Nuraly Akimbekov, PhD (Al-Farabi Kazakh National University, Kazakhstan) for drawing the illustration. Some of the information has been collected from online sources, including OpenAI GPT-3.5 and Google Scholar. The author is solely responsible for the chapter's content. M.S.R. is a visiting professor at the University Kuala Lumpur Royal College of Medicine Perak, Ipoh, Malaysia.

References

1. Razzaque MS, Wimalawansa SJ (2025) Minerals and human health: from deficiency to toxicity. Nutrients 17(3): 454; https://doi.org/10.3390/nu17030454
2. Razzaque MS (2022) Interactions between FGF23 and vitamin D. Endocr Connect 11(10):e220239; https://doi.org/10.1530/EC-22-0239
3. Uwitonze AM, Rahman S, Ojeh N, Grant WB, Kaur H, Haq A, Razzaque MS (2020) Oral manifestations of magnesium and vitamin D inadequacy. J Steroid Biochem Mol Biol 200:105636

4. Erem S, Atfi A, Razzaque MS (2019) Anabolic effects of vitamin D and magnesium in aging bone. J Steroid Biochem Mol Biol 193:105400
5. Uwitonze AM, Razzaque MS (2018) Role of magnesium in Vitamin D activation and function. J Am Osteopath Assoc 118:181–189
6. Zaman MS, Ullah MI (2025) Vitamin D and fracture-healing. In: Razzaque MS (ed) Vitamin D function. Advances in experimental medicine and biology, vol 1493. Springer, Cham, pp 103–111
7. Fischer V, Haffner-Luntzer M, Amling M, Ignatius A (2018) Calcium and vitamin D in bone fracture healing and post-traumatic bone turnover. Eur Cell Mater 35: 365–385
8. Yoshida T, Stern PH (2012) How vitamin D works on bone. Endocrinol Metab Clin North Am 41:557–569
9. Tamanna S (2025) Vitamin D and sleep disorders. In: Razzaque MS (ed) Vitamin D function. Advances in experimental medicine and biology, vol 1493. Springer, Cham, pp 91–100
10. Huiberts LM, Smolders K (2021) Effects of vitamin D on mood and sleep in the healthy population: interpretations from the serotonergic pathway. Sleep Med Rev 55:101379
11. Prono F, Bernardi K, Ferri R, Bruni O (2022) The role of Vitamin D in sleep disorders of children and adolescents: a systematic review. Int J Mol Sci 23
12. Fontes-Dantas RL, Cardoso CR, Domingues V, Fernandes GG (2025) Vitamin D and autoimmunity. In: Razzaque MS (ed) Vitamin D function. Advances in experimental medicine and biology, vol 1493. Springer, Cham, pp 33–44
13. Aranow C (2011) Vitamin D and the immune system. J Investig Med 59:881–886
14. Ahmad R, Chowdhury K, Sinha S, Haque M (2025) Vitamin D and brain health. In: Razzaque MS (ed) Vitamin D function. Advances in experimental medicine and biology, vol 1493. Springer, Cham, pp 51–57
15. Mumena CH, Akimbekov NS, Sasi R, Mlawa M, Mudhihiri MH (2025) Vitamin D and oral cancer. In: Razzaque MS (ed) Vitamin D function. Advances in experimental medicine and biology, vol 1493. Springer, Cham, pp 115–125
16. Jin T, Guo Y, Huang Z, Zhang Q, Huang Z, Zhang Y, Huang Z (2020) Vitamin D inhibits the proliferation of Oral Squamous Cell Carcinoma by suppressing lncRNA LUCAT1 through the MAPK pathway. J Cancer 11:5971–5981
17. Verma A, Vincent-Chong VK, DeJong H, Hershberger PA, Seshadri M (2020) Impact of dietary vitamin D on initiation and progression of oral cancer. J Steroid Biochem Mol Biol 199:105603
18. Ahmed HS (2025) Vitamin D and mineral ion regulation. In: Razzaque MS (ed) Vitamin D function. Advances in experimental medicine and biology, vol 1493. Springer, Cham, pp 69–85
19. Razzaque MS (2009) The FGF23-Klotho axis: endocrine regulation of phosphate homeostasis. Nat Rev Endocrinol 5:611–619
20. Razzaque MS, Mohammadi M (2025) Asymmetric FGF receptor dimerization: implications for FGF23 biology and drug discovery. Am J Physiol Cell Physiol. 329(3):C904-C910.
21. Lanehart D, Uzoma C, Zaman MS, Akimbekov NS, Lopez-Alvarenga JC, Razzaque MS (2025) Maternal-fetal axis: magnesium and pregnancy. In: Razzaque MS (ed) Vitamin D function. Advances in experimental medicine and biology, vol 1493. Springer, Cham, pp 133–138
22. Foamkom A-I, Abdelhady HG, Razzaque MS (2025) Gut-glucose axis: glucagon-like peptide and calcium. In: Razzaque MS (ed) Vitamin D function. Advances in experimental medicine and biology, vol 1493. Springer, Cham, pp 127–130
23. Akimbekov NS, Digel I, Tastambek K, Rodriguez-Raecke R, Tepecik A, Kistaubayeva AS, Zha J, Razzaque MS (2025) Combating bacterial infections with vitamin D-induced antimicrobial peptides. In: Razzaque MS (ed) Vitamin D function. Advances in experimental medicine and biology, vol 1493. Springer, Cham, pp 13–25
24. Akimbekov NS, Digel I, Tastambek K, Rodriguez-Raecke R, Tepecik A, Kistaubayeva AS, Zha J, Razzaque MS (2025) Vitamin D-induced antimicrobial peptides in combating viral infections. In: Razzaque MS (ed) Vitamin D function. Advances in experimental medicine and biology, vol 1493. Springer, Cham, pp 141–150
25. Vasil N, Razzaque MS (2025) Decoding Vitamin D intake: how much is too much? In: Razzaque MS (ed) Vitamin D function. Advances in experimental medicine and biology, vol 1493. Springer, Cham, pp 5–11
26. Razzaque MS (2018) Can adverse effects of excessive Vitamin D supplementation occur without developing hypervitaminosis D? J Steroid Biochem Mol Biol 180: 81–86
27. Lanske B, Razzaque MS (2007) Vitamin D and aging: old concepts and new insights. J Nutr Biochem 18: 771–777
28. Bouillon R, Manousaki D, Rosen C, Trajanoska K, Rivadeneira F, Richards JB (2022) The health effects of Vitamin D supplementation: evidence from human studies. Nat Rev Endocrinol 18:96–110
29. Demay MB, Pittas AG, Bikle DD, Diab DL, Kiely ME, Lazaretti-Castro M, Lips P, Mitchell DM, Murad MH, Powers S, Rao SD, Scragg R, Tayek JA, Valent AM, Walsh JME, McCartney CR (2024) Vitamin D for the prevention of disease: an endocrine society clinical practice guideline. J Clin Endocrinol Metab 109:1907–1947
30. Lips P (2010) Worldwide status of Vitamin D nutrition. J Steroid Biochem Mol Biol 121:297–300
31. Woodford HJ, Barrett S, Pattman S (2018) Vitamin D: too much testing and treating? Clin Med (Lond) 18: 196–200

Decoding Vitamin D Intake: How Much Is Too Much?

Nickolas Vasil and Mohammed S. Razzaque

Abstract

Vitamin D has long been accepted as a vital nutrient for bone health, with past guidelines supporting supplementation to maintain optimal levels. However, the recently updated 2024 Endocrine Society guidelines introduce a more conservative approach, shifting the focus away from widespread supplementation in healthy individuals and instead emphasizing use in high-risk populations. This chapter analyzes the key changes from the previous 2011 guidelines, including the removal of strict serum level thresholds for deficiency, revised daily intake recommendations, and screening criteria. The shift reflects growing evidence suggesting that excessive vitamin D supplementation may not provide additional health benefits and can contribute to risks such as hypercalcemia, kidney dysfunction, and eventual increased healthcare costs. The commentary explains the rationale behind said changes and the value for using evidence-based supplementation. Finally, we discuss the implications of these updated guidelines on paving clinicians' decision-making for their patients while highlighting the need for increased awareness and adherence to ensure safe and effective vitamin D consumption.

Keywords

Vitamin D · Supplementation · Vitamin D guidelines · Vitamin D toxicity · Calcification

Introduction

For decades, vitamin D has been hailed as a vital nutrient for maintaining bone health, with recommendations evolving to encourage widespread supplementation with varying amounts of vitamin D based on an individual's current health status. However, the recently released Endocrine Society guidelines mark a significant shift in literature, cautioning against unnecessary vitamin D intake in otherwise healthy individuals. Vitamin D has many important functions within the body. These include increasing calcium and phosphate absorption within the gastrointestinal tract, reabsorption within the kidneys, decreasing parathyroid hormone secretion, increasing the mineralization, and remodeling of bone. Not only can it aid in maintaining calcium homeostasis, but it may also have numerous immune boosting benefits. These may include modulating inflammatory pathways, promoting the activity of macrophages, reducing pro-inflammatory

N. Vasil
Lake Erie College of Osteopathic Medicine (LECOM), Erie, PA, USA

M. S. Razzaque (✉)
Department of Medical Education, School of Medicine, University of Texas Rio Grande Valley (UTRGV), Edinburg, TX, USA
e-mail: mohammed.razzaque@utrgv.edu

M. S. Razzaque (ed.), *Vitamin D Function*, Advances in Experimental Medicine and Biology 1493,
https://doi.org/10.1007/978-3-032-04357-3_2

cytokines, and helping reduce the risk of certain autoimmune diseases such as multiple sclerosis or rheumatoid arthritis, although there is not enough robust evidence for these possible immunologic benefits at this time to be definitive [1, 2]. Maintaining an adequate level of vitamin D is crucial for all these processes to work efficiently. Numerous sources of vitamin D can be used to secure optimal vitamin D balance within the body. There are two primary forms that can be obtained from the diet: D2 (ergocalciferol) and D3 (cholecalciferol). D2 can be found in plants, fungi, and yeast, while D3 is primarily found in dairy products that have been fortified, such as orange juice and fatty fish, and is synthesized from 7-dehydrocholesterol in the presence of UV light. Once either form is absorbed, it can be hydroxylated in the liver as 25-hydroxychalciferol [25(OH) D] and can be hydroxylated for a second time by the kidney to become the active form, 1,25-dihydroxychalciferol [1,25(OH)$_2$D, calcitriol] (Fig. 1). All the previously mentioned enzymatic steps cannot be completed without the assistance of magnesium as a cofactor [3–5]. Not only does magnesium have a role as a cofactor, but it is necessary to attain the beneficial effects of vitamin D due to the bioactivity of it being a magnesium-dependent process [6, 7]. This is why it is important to recall with any newly released guidelines regarding nutritional intake, such as the Endocrine Society's, that the body may require other nutrients to be able to fully reap the benefits of others such as the magnesium-vitamin D interactions.

Maintaining a healthy intake is important because without ample levels of vitamin D, pathologies such as osteomalacia or rickets may arise. The new guidelines from the Endocrine Society that have been released in July 2024 replaced the old guidelines, which were in place for over 10 years beginning in December 2011. There has been a significant change in the approach of how patients should take supplements. The primary objective of this commentary is to explain and decode what the new changes are, why these changes are necessary, and how they can lead to the prevention of overuse and abuse of vitamin D supplements. The consistent difference between the two published guidelines is that the recommended supplementation has decreased in many clinical scenarios when compared to the earlier report. This begs the question of how much vitamin D may be too much?

Key Changes in the Guidelines

The newly released 2024 guidelines from the Endocrine Society significantly changed the landscape surrounding vitamin D recommendations, including revised daily intake limits and updated thresholds for diagnosing serum status levels, with the aim to provide more evidence-based guidance on appropriate supplementation and testing (Table 1). The definitive thresholds that were previously in place to define one's vitamin D levels as sufficient, insufficient, or deficient have all changed, which will alter the landscape going forward for these previously upheld precise values. Based on the 2011 guidelines, the criteria for various levels were as follows: Levels that were >30 ng/mL indicated sufficiency, while levels at 21–29 ng/mL and <20 ng/mL indicated insufficiency and deficiency, respectively. Within the updated guidelines, there is no longer a specific value that the experts directed at to specify what would be labeled as insufficient, deficient, or sufficient [8, 9]. They left numerical scoring behind due to an uncertainty of evidence that has not shown whether previous criteria are directly linked to health benefits. Along with newly updated criteria values, there have also been updates on recommended dosages for daily supplementation. Part of the aim of these updates may have been put in place to help avoid the risk of hypercalcemia in those who were taking excess, along with the lack of evidence suggesting that dosages past the dietary reference intake (DRI) are beneficial. There are many examples of these conservative changes such as, for those who previously fell into the deficient category which was noted as <20 ng/mL, the previous recommendation was to treat the deficiency with a large loading dose of 50,000 IU/

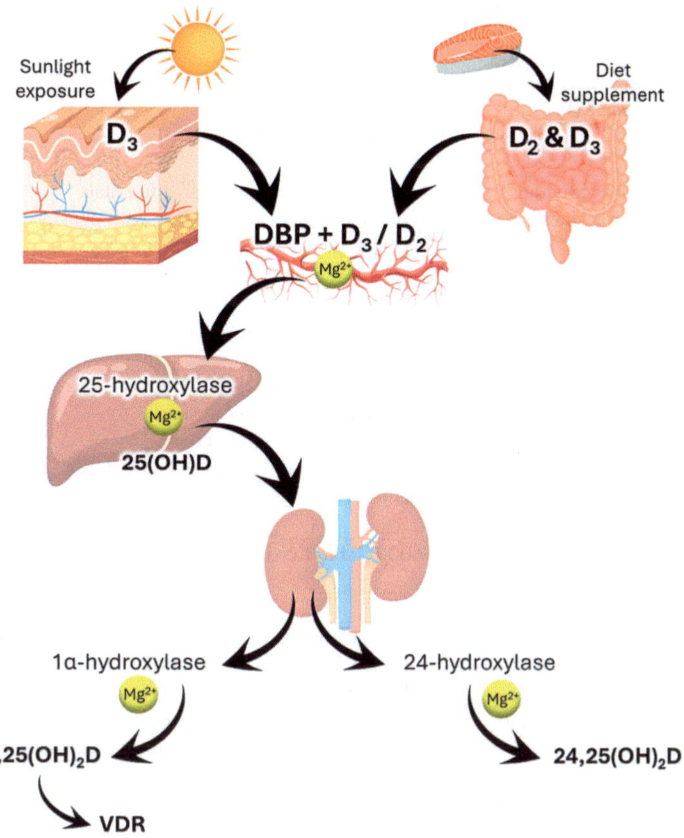

Fig. 1 Vitamin D synthesis along with the role of magnesium in the activation and deactivation of it. Abbreviations: **D3**, vitamin D3 (from animal sources); **D2**, vitamin D2 (nonanimal sources); **DBP**, vitamin D-binding protein; **25(OH)D**, calcifediol (inactive form); **1,25 $(OH)_2D$**, 1,25-dihydroxyvitamin D (biologically active form); **$24,25(OH)_2D$**, 24,25-dihydroxyvitamin D; **Mg**, magnesium; **VDR**, vitamin D receptors

week for 8 weeks. Now for empiric supplementation, lower doses such as 900 IU/day are preferred over the previous intermittent high dosing, and there has not been any mention of such high doses in the new guidelines. In line with the previously mentioned changes, the recommended indications for screening patients for these levels have also changed. The 2011 guidelines stated that "We recommend screening for vitamin D deficiency in individuals at risk for deficiency" [9] which included adults and children with obesity, pregnant individuals, and individuals with dark complexion. Whereas in the 2024 guidelines, "The panel suggests against routine screening for 25(OH)D levels in healthy adults. Screening is only recommended for individuals with clinical symptoms or conditions associated with vitamin D deficiency" [8] which no longer includes healthy individuals with dark complexion, adults with obesity, or pregnant individuals. This may be in part due to the findings that the serum 25(OH)D assay that is commonly used has been found to not be very sensitive or accurate when compared to using liquid chromatography tandem mass spectrometry to measure and calculate the vitamin D metabolite ratio which is not as widely available and may cost more [2, 10]. Following this information, the criteria for screening patients' levels of vitamin D have changed from anyone who is in the at-risk category to only a smaller subset of patients who present with clinical evidence for testing. Overall, there has been a more conservative approach to the new empiric recommendations when comparing the current to the past guidelines. Numerous special patient populations who fall into an "at-risk" category have had their specific guidelines significantly altered within the new release, which is critical to providing the best care possible for each unique patient. Infants' and children's recommendations have been slightly altered from the original doses of 400 IU/d for those under 1 year of age and at

Table 1 Key differences between 2011 and 2024 Endocrine Society vitamin D guidelines [24]

Category	2011 guidelines	2024 guidelines
Serum vitamin D classification	25(OH)D levels Deficiency: <20 ng/mL Insufficiency: 21–29 ng/mL Sufficient: >29 ng/mL	No longer endorses specific 25(OH)D level to define vitamin D sufficiency, insufficiency, and deficiency
Defining individuals at risk for vitamin D deficiency	Suggests screening in high-risk individuals which includes adults and children with obesity and dark complexion or during pregnancy	Suggests against routine 25(OH)D screening in adults and children with obesity and dark complexion or during pregnancy
Criteria for suggested vitamin D screening	Recommended screening for vitamin D deficiency in individuals at risk for deficiency	Only those who present with clinical symptoms
Common dosage recommendations	Recommends maintain 25(OH)D levels >30 ng/mL, often requiring higher dosages (1500–2000 IU/day for adults)	Align intake with dietary reference intake (600–800 IU/day) while discouraging excessive supplementation
Vitamin D treatment dosing for indicated adults	50,000 IU of vitamin D2 or D3 1x/week for 8 weeks or 6000IU daily	Suggests daily, lower-dose vitamin D instead of nondaily, higher-dose vitamin D
Children and adolescents (1–18 years)	400–600 IU/d to maximize bone health	Empiric vitamin D supplementation to prevent rickets and potentially lower risk of respiratory tract infections
Adults (18–74 years)	Suggests at least 600 IU/d of vitamin D. May require 1500-2000 IU/d of vitamin D to maintain blood level of >30 ng/mL	Suggests against empiric vitamin D supplementation past the dietary reference intake and no routine testing
Pregnant individuals	Recommends at least 600 IU/d but recognizes up to 1500–2000 IU/d may be needed to maintain blood level of >30 ng/mL	Suggests empiric vitamin D supplementation but against routine 25 (OH)D testing but no specific value is clear
Adults with high-risk prediabetes	This population was not addressed	Suggests empiric vitamin D supplementation on top of lifestyle changes

least 600 IU/d for any child >1 year old to empirical treatment for any child between the ages of 1 and 18. Another change that was noted was the recommended daily intake of vitamin D through various sources for adults under 75 years of age. Previously in the 2011 guidelines, to maintain serum 25(OH)D levels consistently above 30 ng/mL, adults between the ages of 19 and 70 may require vitamin D intake of at least 1500–2000 IU/day [3]. The new guidelines have significantly lowered these volumes by stating that "Adults under 75 years should adhere to the recommended daily allowance (600–800 IU/day) unless clinically indicated. Empiric supplementation beyond these limits is discouraged" [4]. A vital population of patients who always need to be aware of the current recommendations for vitamin intake are pregnant individuals. The old guidelines recommended a minimum dosage of 600 IU/day but admitted that higher doses, such as up to 2000 IU/d, may be necessary to maintain the prior value of vitamin D sufficiency (>30 ng/mL). While in the updated issue, no specific dose was noted for those who are pregnant to supplement with, but what was specified was that there is suggestion against routine 25(OH) D testing in those who are deemed healthy throughout their pregnancy. The doses used in recent trials have ranged from 2600 to 3300 IU/day, and although supplementation is still recommended during pregnancy, it is simply empiric to help reduce complications such as preeclampsia, and intrauterine mortality, and improve neonatal outcomes. The current guidelines also address a topic that was not mentioned previously: the socioeconomic status of patients and how this status may influence treatment and screening options. Studies on socioeconomic status have shown that this is an important factor that cannot be ignored in regard to treating patients. It can affect the choice of treatments given to patients and their outcomes when it comes to adherence to

treatment plans [11]. This factor was not mentioned in the earlier guidelines but was briefly acknowledged within the update. It discusses how the option of testing may not be feasible owing to resource and cost concerns within this population. Disadvantaged patients and those with low socioeconomic status are at an increased risk for low vitamin D levels; therefore, empiric supplementation would be favorable over routine screening prior to supplementation [12]. Overall, these comparisons reflect a significant shift from the prior, more aggressive vitamin D intake and broader testing recommendations in 2011 to a more conservative, targeted, and evidence-based approach in 2024.

Why the Shift Is Necessary

While the role of vitamin D in bone health is well-established, the rising trend of unnecessary and excessive vitamin D supplementation has raised concerns about unintended consequences that may result from improper use, such as the risk of hypercalcemia, potential kidney damage, toxicity, and raising unnecessary healthcare costs. With all these potential negative effects, the release of these updated guidelines is not only timely but essential for ensuring safe and effective use of this vital nutrient. Within the past decade, there has been a surge in vitamin D testing and supplementation. If all the supplementation was clinically indicated, then these practices may be justified, but a large amount of the rise has not been deemed clinically necessary which is reflected in the new guidelines, which recommended less supplementation and testing [13]. The literature has shown that individuals who are found to be vitamin D replete would not benefit from additional supplements from the start [14]. Individuals who are not deficient but continue to take high dose vitamin D supplements may contribute to being at a higher risk of adverse effects that can come with excess intake of supplements. These findings may have led to changes that were noted within the most recent updated guidelines since a major theme that has been found throughout them is to decrease the number of prescriptions for vitamin D supplementation and to focus on lower amounts taken overall. One of the most common results of excess vitamin D intake is its multiple downstream effects on the body which are caused by hypercalcemia [15]. Excess vitamin D can increase calcium reabsorption from the gastrointestinal tract which raises the serum concentration of calcium above normal levels. This may increase the risk of calcium oxalate kidney stone formation, especially among those who may be more genetically susceptible to vitamin D signaling and calcitriol breakdown [16]. In hypercalcemia, the body places more calcium in the genitourinary tract to be excreted in the urine. Citrate is normally found in the urine and binds to free calcium, preventing stone formation. When there is excess calcium in the urine, this leads to citrate being used up and the calcium is free to bind to oxalate and form stones. Stones when precipitated can lead to severe flank-to-groin pain and hematuria. Excess calcium can cause damage to the kidney and lead to renal dysfunction. Hypercalcemia can also cause stone formation and even central nervous system impairment such as depression [17]. Another rare but possible side effect from excess vitamin D is the ability for it to cause an almost paradoxical effect of bone damage and may increase the risk for bone fractures. Most people know that vitamin D aids in fortifying the bones and that drinking milk is a great source to help with this process. Interestingly, if there is excessive vitamin D taken in, this can lead to an increase in calcium resorption from bones which can induce skeletal deformities at toxic doses [18]. Cardiovascular issues, such as arterial calcification due to hypercalcemia which can stem from the excess supplementation, may arise [19]. With the adjusted dose recommendations, this should lessen the frequency of unnecessary doses of vitamin D due to supplementation in those with previous normal levels. Prior to the updated guidelines, some individuals may opt to undergo routine 25(OH)D testing to monitor their vitamin D levels because of a lack of guidelines saying otherwise prior to the recent update. Not only are some of these tests not clinically useful, but they also add another unnecessary economic burden on patients. Healthcare burdens are a significant cost to patients and may lead them to avoid attending routine appointments [20]. The vitamin D supplement market is valued at over a billion dollars each year and flourishes on lack of understanding surrounding the need to supplement [21]. Many individuals believe that the more vitamin D they can obtain, the healthier their state, especially for their bones, which is clearly not necessarily the case. This is why there are always two sides to the story in regard to whether you should supplement it or not. Based upon the release of the new guidelines, it is safe to presume there is not enough scientific justification to continue with the belief that the more is better.

The Path Forward

To prevent unnecessary use, it is imperative that practicing healthcare providers adopt these updated guidelines while also educating the public on the importance of balanced vitamin D intake and the risks associated with excessive supplementation. Rather than encouraging indiscriminate supplementation, the new guidelines have provided clinicians with a roadmap for a more measured and evidence-based approach to vitamin D intake. Increasing awareness among healthcare providers of updated dietary guidelines is essential to reduce the risk of continued overprescription of dosages that were formerly regarded as standard clinical practice. It is easy to fall behind on updated guidelines since providers may not always be as aware of nutritional guidelines and how frequently they are modified. This applies not only to the dosage updates but also to encourage them to test their patient's vitamin D levels when their clinical presentations warrant this with enough evidence. It will take the entire healthcare system to help debunk the current myth that "the more vitamins you take, the better off you are" that a large percentage of the population may still believe due to vitamin D misinformation that can be found in the media [22]. By implementing adjustments to patients' diets, which may benefit from increased vitamin D, instead of relying on supplements may lead to a more beneficial outcome for their overall health when compared to solely relying on supplementing the deficiency with medications. However, this idea is easier said than done, as maintaining a balanced diet continues to be a struggle for a significant percentage of the population in the United States. As these guidelines lean into the more conservative side and are broader compared to the previous ones, it is important to find the perfect balance between benefits and risks given the current data. Based on the guidelines, it is a safe assumption that most patients who are healthy should stick to the recommended DRI level of vitamin D. This allows patients who are trying to follow these recommendations to track their intake of vitamin D in an easier way since daily value percentages, which are based on DRIs, are found on the back of most food labels in the United States. These labels aid in the ability to monitor vitamin D intake throughout the day from the diet. This is a good steppingstone for those who are unsure of where to start. The special populations mentioned above that are encouraged to supplements may find themselves lost in what direction they should go for when it comes to supplementation without the advice from their doctor. These populations are why it is of the utmost importance to raise the awareness of the new guidelines. There is much further research to still be done on vitamin D to help find the optimal dosage thresholds for the nonskeletal benefits that it may provide. Long-term clinical trials will be necessary to determine what levels may aid in the outcomes varying on patient diseases. With the current information that is in the literature, it is vital to make sure we always balance the benefits and risks to the best of our clinical judgement when it comes to advising supplementation. This cannot be done without the advancements of clinical research and staying up to date with constantly updated guideline releases such as the most recent edition discussed here from the Endocrine Society. It is vital to take expert advice, such that it has been elaborated in recent guidelines. Supplementing primarily high-risk groups seems to be the mainstay scenario for supplementing anything past a well-rounded diet.

Spreading education to the masses is an important part for everyone to be able to benefit from the decades of research on this crucial vitamin. In only a few short years, there have been significant changes in recommendations, which is a clear sign that there is still plenty of research to be done within the field. Some of the most important findings are yet to be discovered when it comes to risk-benefit analysis. Optimal dosage thresholds for all the nonskeletal benefits have yet to produce concrete and consistent results. Its role in aiding immune function, cancer prevention, and how it may help those with chronic diseases is still an area of wide variability with the current data, which leaves a significant gap in the literature that needs to be explored. With the current state of advancing technology, more clinically relevant research may lead to an even newer set of guidelines released sooner than what we may expect [23].

Conclusion

The 2024 Endocrine Society guidelines serve as a wake-up call, urging a more rational and conservative approach to vitamin D supplementation in a world where overuse has become widespread and what some may even argue, encouraged. These updated vitamin D guidelines mark an important step toward addressing the overuse of supplements and show how a more balanced approach to supplementation will help ensure the health and safety of individuals globally. Recently updated guidelines on vitamins have shown that there will always be continuing research to help provide the most up to date scientific evidence as to what the optimal levels for these vital vitamins are for every patient along with their current health status which may be implemented with future advancements in precision medicine. As new genomic and metabolic testing options continue to rise, we may be able to push the boundaries of precision medicine, which is highly individualized medical care plans designed for patients by using genetic and metabolic profiling, to include every aspect of health plans including individualized vitamin D supplementation dosages as necessary [24]. Every person is different and may require a different dose, which is why these updates are critical for providing the best treatment possible for patients. As we move forward, the challenge lies not in how much vitamin D we can take, but in how wisely we use it.

Acknowledgment We want to express our sincere gratitude to Dr. Nuraly Akimbekov, PhD (Al-Farabi Kazakh National University, Kazakhstan) for drawing the illustration. N.V. is a Doctor of Osteopathic Medicine (DO) student at the Lake Erie College of Osteopathic Medicine (LECOM), Erie, USA. M.S.R. is a visiting professor at the University Kuala Lumpur Royal College of Medicine Perak, Ipoh, Malaysia.

References

1. Aranow C (2011) Vitamin D and the immune system. J Investig Med 59:881–886
2. Altieri B, Cavalier E, Bhattoa HP, Perez-Lopez FR, Lopez-Baena MT, Perez-Roncero GR, Chedraui P, Annweiler C, Della Casa S, Zelzer S, Herrmann M, Faggiano A, Colao A, Holick MF (2020) Vitamin D testing: advantages and limits of the current assays. Eur J Clin Nutr 74:231–247
3. Erem S, Atfi A, Razzaque MS (2019) Anabolic effects of vitamin D and magnesium in aging bone. J Steroid Biochem Mol Biol 193:105400
4. Razzaque MS (2018) Magnesium: are we consuming enough? Nutrients 10:1863
5. Uwitonze AM, Rahman S, Ojeh N, Grant WB, Kaur H, Haq A, Razzaque MS (2020) Oral manifestations of magnesium and vitamin D inadequacy. J Steroid Biochem Mol Biol 200:105636
6. Eskander M, Razzaque MS (2022) Can maintaining optimal magnesium balance reduce the disease severity of COVID-19 patients? Front Endocrinol (Lausanne) 13:843152
7. Uwitonze AM, Razzaque MS (2018) Role of magnesium in Vitamin D activation and function. J Am Osteopath Assoc 118:181–189
8. Demay MB, Pittas AG, Bikle DD, Diab DL, Kiely ME, Lazaretti-Castro M, Lips P, Mitchell DM, Murad MH, Powers S, Rao SD, Scragg R, Tayek JA, Valent AM, Walsh JME, McCartney CR (2024) Vitamin D for the prevention of disease: an endocrine society clinical practice guideline. J Clin Endocrinol Metab 109:1907–1947
9. Holick MF, Binkley NC, Bischoff-Ferrari HA, Gordon CM, Hanley DA, Heaney RP, Murad MH, Weaver CM, Endocrine S (2011) Evaluation, treatment, and prevention of vitamin D deficiency: an Endocrine Society clinical practice guideline. J Clin Endocrinol Metab 96:1911–1930

10. Carter GD (2011) Accuracy of 25-hydroxyvitamin D assays: confronting the issues. Curr Drug Targets 12: 19–28
11. Bernheim SM, Ross JS, Krumholz HM, Bradley EH (2008) Influence of patients' socioeconomic status on clinical management decisions: a qualitative study. Ann Fam Med 6:53–59
12. Lips P (2010) Worldwide status of vitamin D nutrition. J Steroid Biochem Mol Biol 121:297–300
13. Woodford HJ, Barrett S, Pattman S (2018) Vitamin D: too much testing and treating? Clin Med (Lond) 18: 196–200
14. Bouillon R, Manousaki D, Rosen C, Trajanoska K, Rivadeneira F, Richards JB (2022) The health effects of vitamin D supplementation: evidence from human studies. Nat Rev Endocrinol 18:96–110
15. Marcinowska-Suchowierska E, Kupisz-Urbanska M, Lukaszkiewicz J, Pludowski P, Jones G (2018) Vitamin D toxicity – a clinical perspective. Front Endocrinol (Lausanne) 9:550
16. Letavernier E, Daudon M (2018) Vitamin D, hypercalciuria and kidney stones. Nutrients 10:366
17. Joborn C, Hetta J, Lind L, Rastad J, Akerstrom G, Ljunghall S (1989) Self-rated psychiatric symptoms in patients operated on because of primary hyperparathyroidism and in patients with long-standing mild hypercalcemia. Surgery 105:72–78
18. Suda T, Takahashi F, Takahashi N (2012) Bone effects of vitamin D – discrepancies between in vivo and in vitro studies. Arch Biochem Biophys 523:22–29
19. Carroll MF, Schade DS (2003) A practical approach to hypercalcemia. Am Fam Physician 67:1959–1966
20. Iuga AO, McGuire MJ (2014) Adherence and health care costs. Risk Manag Healthc Policy 7:35–44
21. Lordan R (2021) Notable developments for Vitamin D amid the COVID-19 pandemic, but caution warranted overall: a narrative review. Nutrients 13:740
22. O'Connor C, O'Grady C, Murphy M (2022) Spotting fake news: a qualitative review of misinformation and conspiracy theories in acne vulgaris. Clin Exp Dermatol 47:1707–1711
23. Chopra H, Annu SDK, Munjal K, Priyanka, Dhama K, Emran TB (2023) Revolutionizing clinical trials: the role of AI in accelerating medical breakthroughs. Int J Surg 109:4211–4220
24. Liu S, Liu Q (2018) Personalized magnesium intervention to improve vitamin D metabolism: applying a systems approach for precision nutrition in large randomized trials of diverse populations. Am J Clin Nutr 108:1159–1161

Combating Bacterial Infections with Vitamin D-Induced Antimicrobial Peptides

Nuraly S. Akimbekov, Ilya Digel, Kuanysh Tastambek, Rea Rodriguez-Raecke, Atakan Tepecik, Aida S. Kistaubayeva, Jian Zha, and Mohammed S. Razzaque

Abstract

Owing to emerging global threats such as viral pandemics and the rise of antibiotic resistance, there is an urgent need for novel approaches to combat drug-resistant pathogens. Antimicrobial peptides (AMPs), also known as cationic host defense peptides (CHDPs), are considered promising candidates because of their ability to target various pathogenic microorganisms both directly and indirectly. These functionally versatile molecules modulate host immune responses by activating multiple signaling pathways simultaneously. The unique mechanisms of interaction within the pathogen-host system, combined with relatively straightforward biosynthesis, make AMPs viable alternatives to traditional antibiotics. Vitamin D is a key endogenous factor that regulates the biosynthesis and activity of AMPs, particularly cathelicidin and defensins. This chapter focuses on the interplay between vitamin D and antimicrobial peptides, highlighting recent advances in understanding their synergistic protective effects and potential clinical benefits.

N. S. Akimbekov (✉)
International Center for Islamic Science and Innovation, Al-Farabi Kazakh National University, Almaty, Kazakhstan

Scientific-Practical Center, West Kazakhstan Marat Ospanov Medical University, Aktobe, Kazakhstan

Ecology Research Institute, Khoja Akhmet Yassawi International Kazakh-Turkish University, Turkistan, Kazakhstan
e-mail: Akimbekov.Nuraly@kaznu.edu.kz

I. Digel · A. Tepecik
Institute for Bioengineering, Aachen University of Applied Sciences, Jülich, Germany

K. Tastambek
International Center for Islamic Science and Innovation, Al-Farabi Kazakh National University, Almaty, Kazakhstan

Ecology Research Institute, Khoja Akhmet Yassawi International Kazakh-Turkish University, Turkistan, Kazakhstan

R. Rodriguez-Raecke
Interdisciplinary Center for Clinical Research (IZKF), RWTH Aachen University, Aachen, Germany

A. S. Kistaubayeva
International Center for Islamic Science and Innovation, Al-Farabi Kazakh National University, Almaty, Kazakhstan

J. Zha
School of Food and Biological Engineering, Shaanxi University of Science and Technology, ShaanxiXi'an, China

M. S. Razzaque
University Kuala Lumpur Royal College of Medicine Perak (UniKL RCMP), Ipoh, Perak, Malaysia

Department of Medical Education, School of Medicine, University of Texas Rio Grande Valley (UTRGV), Edinburg, TX, USA

M. S. Razzaque (ed.), *Vitamin D Function*, Advances in Experimental Medicine and Biology 1493,
https://doi.org/10.1007/978-3-032-04357-3_3

Keywords

Antimicrobial peptide · Cathelicidin · Defensins · Infection · Innate immunity · Vitamin D · Vitamin D receptor

Introduction

Decades of research have shown that maintaining optimal vitamin D levels is crucial for the proper functioning of most physiological systems in the human body [1–4]. Despite this knowledge, vitamin D deficiency remains a significant global health issue. In addition to its well-known role in regulating body calcium levels, vitamin D, particularly in its active forms, such as 1,25 $(OH)_2D$, plays a vital role in innate immunity [5, 6]. In recent years, vitamin D has increasingly been recognized as a complex regulator of both innate and adaptive immune responses, with increasing evidence suggesting that vitamin D deficiency impairs antibacterial and antiviral immunity [7]. Moreover, the involvement of vitamin D in inflammatory processes and its influence on infection risk reduction have been well documented [8, 9].

Vitamin D receptors (VDRs) are expressed in many immune cells, making them sensitive to vitamin D-mediated induction [10]. In addition to modulating immune cell growth and differentiation, $1,25(OH)_2D$ regulates the expression of genes encoding antimicrobial peptides (AMPs) [11–16]. The measured levels of AMPs strongly positively correlate with the availability of vitamin D in patients with sepsis [17–19], pelvic inflammatory disease [20], cystic fibrosis [21], urinary tract infections [22, 23], inflammatory bowel disease [24, 25], leprosy (Hansen's disease) [26, 27], periodontal diseases [28, 29], and many other disorders. Extensive clinical studies conducted between 2014 and 2016 revealed strong links between vitamin D deficiency, chronic inflammation, and disorders of innate immunity, which can manifest as a significant reduction in the body's resistance to tuberculosis and chronic bronchitis [7, 30].

This chapter provides a brief overview of AMPs, outlining their characteristics and mechanisms of antimicrobial action. It also describes known interactions between vitamin D and AMPs, highlighting the molecular basis for the antimicrobial effects associated with vitamin D. Additionally, we discuss relevant findings from both laboratory and clinical studies regarding the effectiveness of vitamin D against bacterial infections.

Characteristics of AMPs

AMPs, also known as "natural antibiotics," are an evolutionarily old component of the innate immune system in most multicellular organisms [31]. AMPs consist of 10–50 amino acid residues and have a molecular weight of less than 10 kDa [32, 33]. The primary structure of these cationic peptides features over 30% hydrophobic amino acid residues [34], which presumably facilitates the targeting of negatively charged microbial membranes [35]. In mammals, two structurally and biologically distinct families of broad-spectrum AMPs have been identified: *cathelicidins* and *defensins* (α- and β-). The only cathelicidin expressed in the human body is hCAP-18, which is expressed mostly in epithelial cells and neutrophils. The product of its cleavage, LL-37, has a range of biological activities, including lipopolysaccharide (LPS) neutralization, direct antibacterial effects, and the modulation of the inflammatory response [36].

The main features of AMPs are summarized in Table 1.

The Antibacterial Activity of AMPs

AMPs exhibit a broad spectrum of activity against bacteria, viruses, fungi, and unicellular protozoa. This occurs through various mechanisms, primarily by targeting and destroying the essential components of bacterial cells that are necessary for their survival. AMPs can interact with different elements of microbial

Table 1 General characteristics of AMPs

Property	Cathelicidins	Defensins
Genes	Approx. 2 kb, located on chromosome 3 (band 3p21.31). Four-exon/three-intron organization [37, 38]	Clustered on chromosome 8 (band 8p22-23). Only α- and β-defensins are expressed in humans. θ-defensins have been identified thus far only in leukocytes of the rhesus macaque [39]
Structure	Linear peptides <10 kDa with 23–37 amino acid residues fold into amphipathic α-helices. Some cathelicidins may possess beta-hairpin structures, stabilized by disulfide bonds. Larger cathelicidins display repetitive proline motifs	Small (approx. 4 kDa) cysteine-rich cationic peptides, typically 18–45 amino acid residues. Triple-stranded antiparallel beta sheets are stabilized by three to four intramolecular disulfide bridges [40, 41]. The S-S bonds confer resistance to proteinases and thus ensure the stability of antimicrobial properties at the inflammation site [42]
Biosynthesis	Epithelial cells, keratinocytes, mast cells, NK cells, dendritic cells, macrophages, neutrophils, lymphocytes, and some other cells [43, 44] To date, the only member of the cathelicidin family found in humans is hCAP-18 (cleaved into LL-37 and FALL-39) [45]	Human α-defensins HNP-1 to HNP-4 are expressed by neutrophils, immature dendritic cells, monocytes, macrophages, NK cells, T lymphocytes, and epithelial cells of the cervix, placenta, and intestine [46]. A large amount of pro-HNPs is synthesized in myelocytes HD-5 and HD-6 are produced by intestinal Paneth cells, vaginal epithelial cells, salivary glands, and epithelium of the digestive and urinary systems and in the mucous membrane of the eyes [47] Human β-defensins HBD-1 to HBD-4 are expressed in various tissues, such as the trachea, bronchus, mammary gland, and intestinal epithelial cells [48]
Inducer	Pro-inflammatory cytokines, growth factors, 1,25 $(OH)_2D$ [49]	Pro-inflammatory cytokines, growth factors, 1,25 $(OH)_2D$ [50]

cells, such as membranes, cell walls, DNA, protein synthesis machinery, cytokines, enzymes, and reactive oxygen species, allowing them to effectively combat a wide range of pathogens (Fig. 1).

The microbial *cell wall* is a major target of AMPs. Most of the antibacterial effects of AMPs arise from their ability to interfere with the synthesis of cell wall components and to damage the cell wall structure. The specific mechanisms by which AMPs affect the cell wall include the following [51, 52]:

- Interfering with the biosynthesis of peptidoglycan, a key component of the cell wall, which causes damage to its structure
- Compromising cell wall integrity by inhibiting the biosynthesis of wall teichoic acid
- Activating autolysins, such as Pep5 and nisin
- Inducing the aggregation of LPS micelles in gram-negative bacteria

The primary mechanism of antibacterial action for most AMPs involves direct disruption of the *cell membrane*, which is essential for maintaining cellular integrity. Differences between bacterial and mammalian cell membranes allow AMPs to selectively target and damage microbial cells while sparing host cells. Both gram-positive and gram-negative bacteria share fundamental similarities in the composition of their cytoplasmic membranes. Their membranes are rich in phosphatidylglycerol and cardiolipin, which are characterized predominantly by anionic head groups. These negatively charged components create a significant electrostatic attraction for cationic AMPs, allowing them to interact specifically with bacterial membranes [53].

Additionally, AMPs interact with the phospholipid bilayer through hydrophobic amino acid residues, binding to the membrane surface to form an initial membrane-bound complex. Upon binding, AMPs undergo conformational

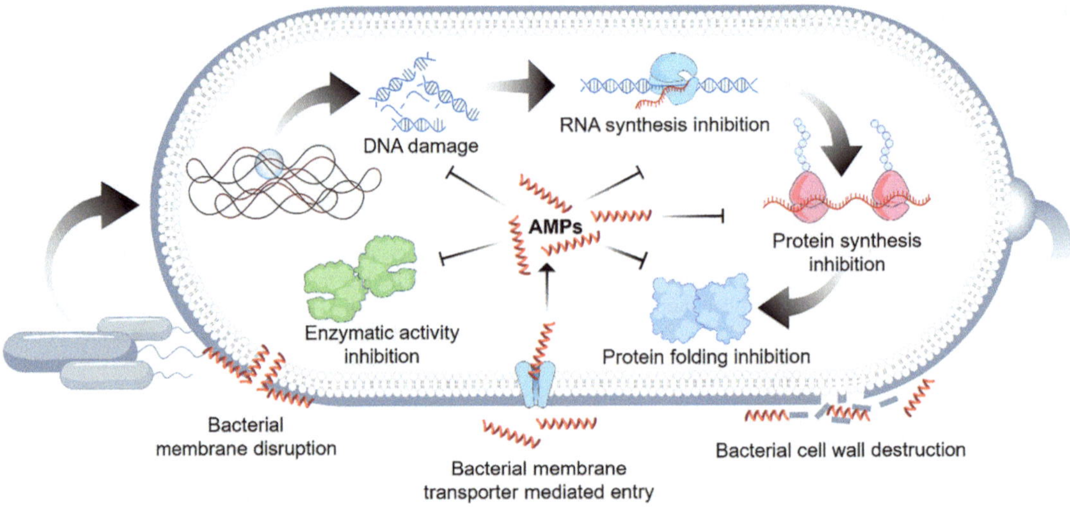

Fig. 1 Antibacterial mechanisms of AMPs. AMPs induce bacterial cell death through multiple mechanisms, including cell wall targeting, membrane disruption, and interference with major metabolic processes, such as damage to DNA, inhibition of RNA and protein synthesis, disturbance of protein folding, and reduced enzymatic activity

changes, transitioning from a disordered or "loose" state into a more organized secondary structure (α-helices or β-sheets). As AMPs integrate into the membrane, they cause local disruption, leading to the formation of transmembrane pores that cause uncontrolled leakage of small molecules and ions. Continued insertion and pore formation by AMPs compromise the integrity of the cell membrane, ultimately leading to its breakdown [54].

Several mechanisms of cytoplasmic bacterial membrane disruption by AMPs have been proposed independently by Wimley et al., Chang et al., and Zhang et al., as illustrated in Fig. 2. All of these models are based on membrane permeabilization and include the "barrel-stave," "toroidal pore," "carpet-like," and "detergent-like" models [55–57].

According to the "barrel-stave" model, the formation of pores in membranes occurs via the lateral diffusion of peptides through the lipid bilayer. AMPs form a membrane-trespassing spiral shape, in which their hydrophobic sites bind to the lipid bilayer, whereas hydrophilic sites bind together [58]. The "toroidal pore" model is based on the observation that AMPs typically align perpendicularly to the membrane. In this case, AMP molecules interact with the lipid structures of the bacterial membrane, creating toroidal pores [59].

In the "carpet-like" model, the AMPs first accumulate and orient themselves parallel to the surface of the target membrane and then electrostatically interact with the anionic groups of phospholipid heads at many sites, covering the membrane in a carpet-like manner. When a critical concentration is reached, pores are formed in the target membrane, leading to bacterial lysis [55]. Finally, the "detergent-like" model assumes that bacterial membrane integrity and pore size-independent leakage collapse when the AMP concentration exceeds a critical value [60].

In addition to influencing the cell wall and membrane of bacterial cells, AMPs can also act on *intercellular targets* through endocytosis or via direct penetration [61]. However, there is still some uncertainty regarding the mechanisms of AMP transmembrane action. Research has shown that AMPs can translocate through membrane defects, traverse cell membranes by creating temporary toroidal gaps, or move via receptor-mediated transport pathways (Fig. 1).

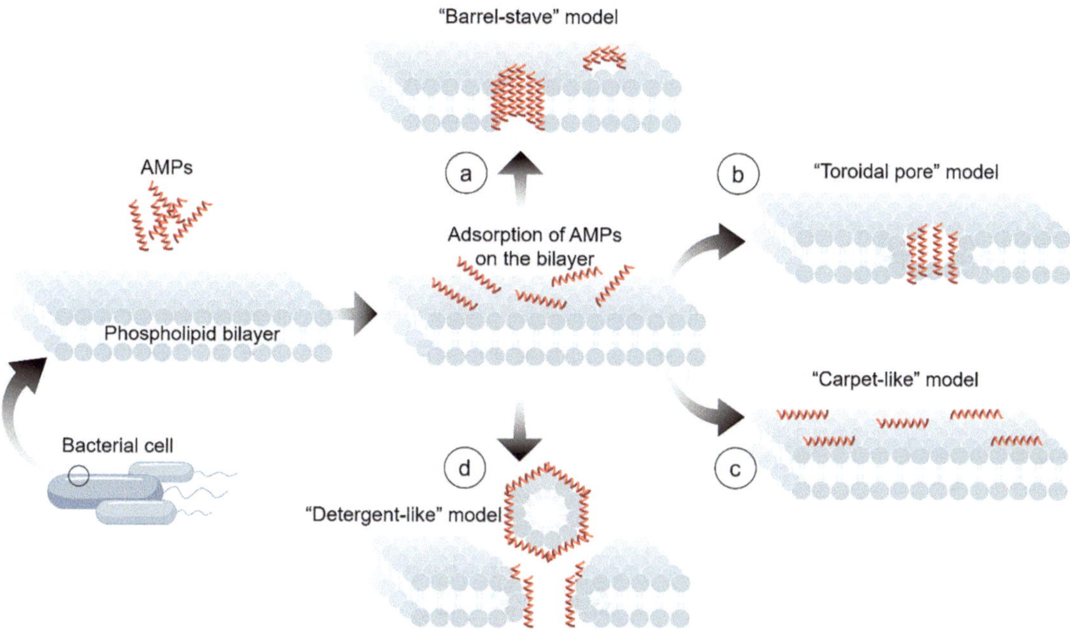

Fig. 2 Models of the bacterial membrane-disrupting mechanisms of AMPs: (**a**) "barrel-stave," (**b**) the "toroidal pore," (**c**) "carpet-like," (**d**) "detergent-like"

Inside the cell, they can target cytoplasmic proteins, nuclei, and organelles, thus exerting antimicrobial effects. There are several principal intracellular target mechanisms of AMPs [62–64]:

- Binding of AMPs to nucleic acids disrupts their structure and prevents the synthesis of DNA, RNA, and proteins.
- Suppression of enzyme and signaling protein activity in metabolic pathways associated with nucleic acid and protein synthesis.
- Destruction of nucleic acid damage repair pathways.
- Ribosome suppression.
- Interference with proper folding and assembly of proteins.
- Inhibition of cell division and blockage of the cell cycle.

Diverse mechanisms of AMP action hinder the development of bacterial resistance and help maintain their effectiveness against a broad spectrum of microorganisms. However, their efficacy can vary across different bacterial species, as summarized in Table 2.

Interactions Between Vitamin D and AMPs

The roles of vitamin D signaling in immunity have been extensively studied, ranging from the stimulation of antibacterial and antiviral responses to the suppression of autoimmunity [22, 109, 110]. A significant finding was that $1,25(OH)_2D$ (calcitriol), the active form of vitamin D, induces the transcription of genes encoding AMPs [111, 112]. The identification of vitamin D response elements (VDREs) near the transcription start sites of the cathelicidin antimicrobial peptide (CAMP/LL37) and β-defensin 2 (DEFB2/DEFB4/HBD2) genes confirmed that these AMPs are direct targets of the VDR [113]. Notably, HBD-2 induction is specific to epithelial cells, whereas CAMP expression is significantly differently induced across various cell types [114]. This regulatory mechanism appears to be specific to humans and primates. Indeed, J.H. White reported that the VDRE in the human CAMP promoter is located within a human-/primate-specific Alu repeat transposable element, which seems to have been introduced before the

Table 2 Antibacterial coverage of AMPs

		Defensins									
Microorganisms	LL-37	HNP				HD		HBD			
		1	2	3	4	5	6	1	2	3	4
Bacteria											
Acinetobacter baumannii	+ [65]					+ [66]			+ [67]	+ [68]	+ [69]
Actinobacillus actinomycetemcomitans										+ [70]	
Bacillus cereus		+ [71]	+ [71]	+ [71]	+ [71]	+ [71]	+ [71]				
Bacillus spp.	+ [65]										
Bacillus subtilis						+ [72]			+ [73]		
Bifidobacterium spp.								+ [74]			
Bifidobacterium adolescentis							+ [75]				
Clostridium difficile		+ [76]				+ [76]					
Enterobacter aerogenes		+ [71]	+ [71]	+ [71]	+ [71]	+ [71]	+ [71]				
Enterococcus faecalis		+ [76, 77]				+ [76]			+ [67, 78]	+ [79]	
Enterococcus faecium									+ [67]	+ [68]	
Enterococcus spp.	+ [65]										
Escherichia coli	+ [65, 80, 81]	+ [71, 76, 80, 82–85]	+ [71, 82, 83, 85]	+ [71, 82, 83]	+ [71, 86]	+ [71, 72, 76, 83]	+ [71, 83]	+ [87]	+ [87–89]	+ [88, 90]	
Haemophilus influenzae	+ [91]							+ [92]	+ [92, 93]		
Klebsiella pneumoniae	+ [65]	+ [94]								+ [94]	
Lactobacillus acidophilus	+ [65]										
Lactobacillus spp.								+ [74]			
Listeria monocytogenes	+ [65, 80]	+ [80]								+ [95]	
Moraxella catarrhalis								+ [92]			
Mycobacterium avium		+ [96]	+ [96]	+ [96]				+ [97]			+ [97]
Mycobacterium intracellulare		+ [96]	+ [96]	+ [96]							
Neisseria gonorrheae	+ [65]										

Propionibacterium acnes	+ [65]										
Proteus vulgaris	+ [65]										
Pseudomonas aeruginosa	+ [65, 80]	+ [80, 82]	+ [82]	+ [82]		+ [72]			+ [67, 89, 98]	+ [70, 88, 99]	+ [69]
Salmonella enterica	+ [100]							+ [85]	+ [85]		
Salmonella typhi								+ [87]	+ [87]		
Salmonella typhimurium	+ [65, 80]	+ [80]	+ [101]	+ [101]			+ [75, 102]				
Staphylococcus aureus	+ [86, 103]	+ [82, 84–86]	+ [67, 71, 82, 85]	+ [71, 82]	+ [71]	+ [71, 72]	+ [71]	+ [104]	+ [88]	+ [68, 88, 105]	+ [69]
Staphylococcus epidermidis	+ [80, 106]	+ [80]									
Staphylococcus spp.	+ [65]										
Stenotrophomonas maltophilia	+ [65]									+ [68]	
Streptococcus faecalis					+ [86]						
Streptococcus gordonii										+ [79]	
Streptococcus mutans									+ [107]	+ [68, 79]	
Streptococcus pneumoniae								+ [92]	+ [92, 108]		
Streptococcus spp.	+ [65]										

evolutionary split between old-world and new-world primates [115]. Similarly, the VDRE in the HBD2 promoter is not conserved in rodents, and no evidence of calcitriol-induced expression of mouse homologs of CAMP or HBD2 has been found in mouse cells of epithelial or myeloid origin. In contrast to human cells, murine cells do not exhibit antibacterial activity when exposed to calcitriol [116].

Monocytes and macrophages play vital roles in the innate immune system. They are capable of both phagocytosing pathogens and sensing pathogen-associated molecular patterns (PAMPs) that these pathogens express. This sensing is conducted through pathogen recognition receptors (PRRs), such as toll-like receptors (TLRs), which are found on various cell types, including monocytes [117, 118]. The functional TLRs identified in humans respond to specific PAMPs from a diverse array of microbes, including bacteria, viruses, protozoa, and fungi [118]. Commonly studied TLRs include TLR2, which responds to gram-positive bacteria and mycobacteria; TLR4, which responds to gram-negative bacteria; and TLR3, which is activated by double-stranded RNA associated with viral infections [117]. Although TLRs are closely linked to the innate responses of monocytes to infection, they have also been identified in a variety of other cell types, including dendritic cells and lymphocytes.

TLRs allow monocytes to perform phagocytosis in response to pathogens (Fig. 3). This action is crucial for minimizing potential damage to host cells by invading bacteria. A study conducted in 2006 by Liu et al. explored the genetic responses activated in monocytes when they sensed *Mycobacterium tuberculosis* via TLR2/TLR1. The findings revealed a specific regulation of genes, particularly CYP27B1 and VDR [119], highlighting the presence of an intracrine system, where locally synthesized 1,25(OH)$_2$D binds to the endogenous VDR, ultimately guiding gene expression within monocytes [120–122]. One significant target of this intracrine response is LL-37, which is directly activated by the 1,25(OH)$_2$D-VDR complex [113]. Functional analyses have shown that after TLR2/TLR1 activation in *M. tuberculosis*, the presence of 1,25(OH)$_2$D enhances LL-37 expression in monocytes and macrophages. This increase has been linked to improved *M. tuberculosis* killing efficacy [119]. Interestingly, this understanding now indicates that the production of LL-37 in monocytes relies on TLR2/TLR1 activation alongside the availability of vitamin D, which supports this vital intracrine mechanism. Further studies have confirmed a correlation between serum vitamin D levels and monocyte LL-37 production [11, 49, 114]. Consequently, in individuals with low vitamin D levels, it may be more challenging to induce LL-37 in monocytes, which leads to increased susceptibility to infections [123]. Notably, vitamin D supplementation in individuals who are deficient has shown promise in enhancing TLR-mediated induction of monocyte LL-37, potentially offering a valuable form of enhancing protection against infections [124, 125].

The vitamin D steroidogenic system is unique because it includes a specific feedback control enzyme called 24-hydroxylase (CYP24A1), which produces less active 24-hydroxylated metabolites from both 25OHD and 1,25(OH)$_2$D. The expression of CYP24A1 is primarily induced by its main substrate, 1,25(OH)$_2$D. In monocytes, locally synthesized 1,25(OH)$_2$D stimulates CYP24A1 while also triggering the intracrine induction of LL-37 [119]. Therefore, CYP24A1 may play a crucial role in determining how monocytes respond to 1,25(OH)$_2$D by converting it into the less active form 1,24,25(OH)$_3$D (1,24,25-trihydroxyvitamin D) [126]. The significance of this process is highlighted by K. Edfeldt et al., who reported that the cytokine IL-4 enhances 24-hydroxylase activity in monocytes, which in turn reduces the TLR2-/TLR1-mediated intracrine induction of LL-37 expression by 1,25(OH)$_2$D [110].

Vitamin D-Dependent AMP Expression in Epithelial Cells

VDR-mediated production of AMPs is also a characteristic feature of epithelial cells of the

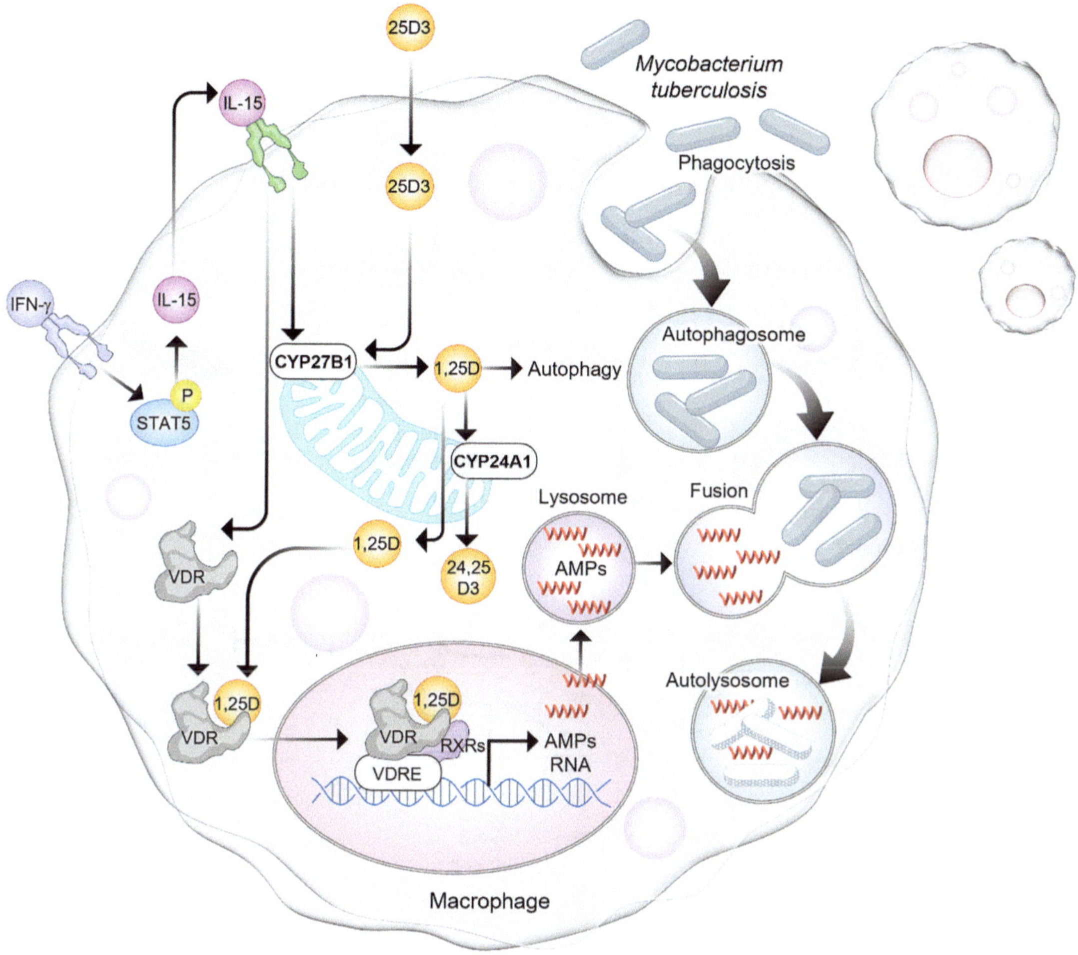

Fig. 3 Vitamin D-induced AMP production in macrophages to target *M. tuberculosis*. Upon entering macrophages, vitamin D is converted into 1,25(OH)$_2$D via CYP27B1. This active form, 1,25(OH)$_2$D, then interacts with the VDR/RXR heterodimer in the cytoplasm. The resulting complex, 1,25(OH)$_2$D/VDR/RXR, translocates to the nucleus, where it binds to specific regulatory elements. This binding modulates the transcription of target genes by promoting the expression of AMPs. The incorporation of AMPs into autophagosomes containing internalized pathogens enables them to function as antibacterial agents

digestive and respiratory tracts. The vitamin D-dependent regulation of AMP production in epithelial cells of the large intestinal mucosa, which regularly interact with commensal intestinal bacteria, differs significantly from VDR-mediated AMP induction by other cells (Fig. 4).

In human epithelial cells, vitamin D/VDR signaling directly stimulates the transcription of AMP-encoding genes. The majority of AMPs are transcribed from CAMP and HBD2. VDRs bind to vitamin D response elements located in the promoter regions of CAMP and HBD2, which are positioned 507 and 1231 base pairs upstream of their respective transcription initiation sites, respectively [113, 127, 128]. Interestingly, this regulatory mechanism, which is observed in human epithelial and myeloid cells upon vitamin D exposure, does not occur in murine cells [116].

Wang et al. [129] reported that the 1,25 (OH)$_2$D signaling pathway is functionally linked to the NOD2 cytoplasmic receptor. The activation

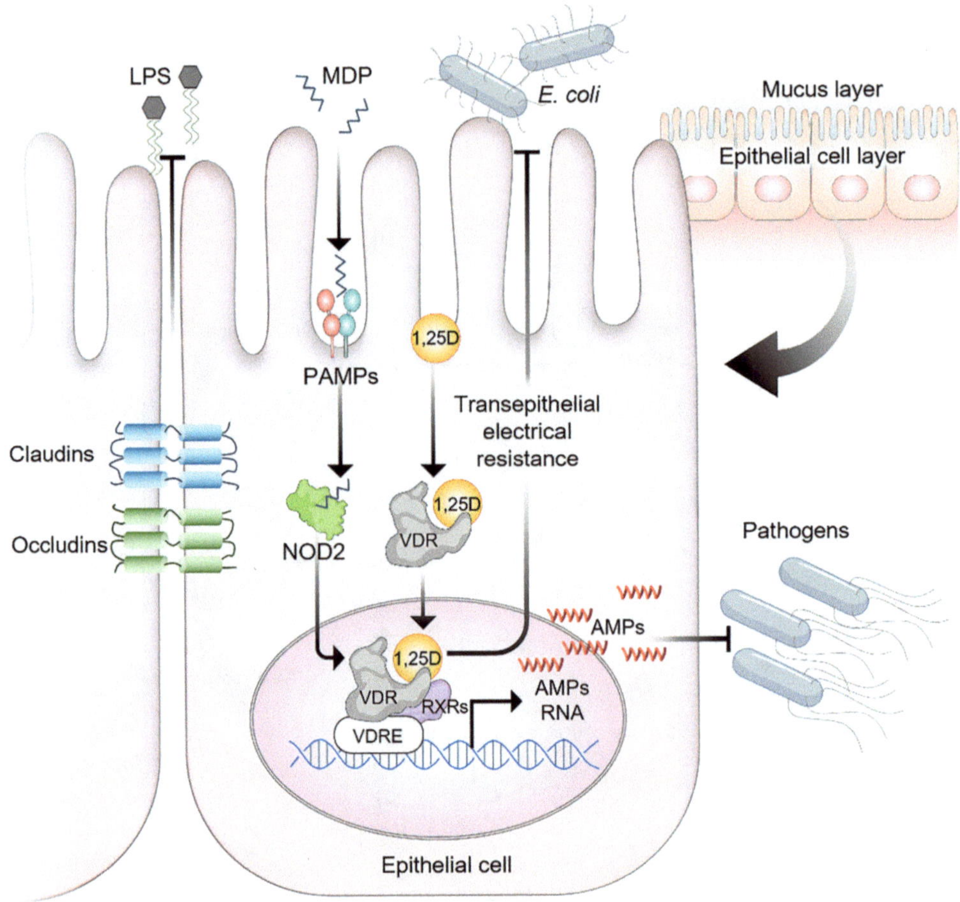

Fig. 4 Involvement of human intestinal epithelial cells in VDR-mediated expression of AMPs. Activated VDRs bind to VDREs to regulate the expression of AMP genes, leading to the production of CAMP and HBD2. In addition, VDRs help maintain intestinal barrier function by preventing LPS translocation. VDREs inhibit the *E. coli*-induced reduction in transepithelial electrical resistance and regulate inflammatory responses by promoting the expression of tight junction proteins (occludins and claudins), thus stabilizing intestinal epithelium permeability

of NOD2 with muramyl dipeptide under 1,25 $(OH)_2D$ synergistically enhances the expression of CAMP and HBD2 in epithelial cells. Research by Zhang et al. demonstrated that VDRs control the expression of tight junction proteins, specifically ZO-1 and ZO-2, by upregulating claudin 2 and 12 and downregulating adherin-17 in the intestinal tract [130]. Moreover, Zhao et al. reported that VDR expression strengthened tight junctions between cells, thereby lowering LPS levels in Caco-2 cells incubated with or without dextran sodium sulfate, resulting in a reduction in intestinal permeability [131]. VDRs can also prevent the reduction in transepithelial electrical resistance caused by pathogenic *E. coli* O157: H7, thus lowering the permeability of the intestinal barrier [132].

Vitamin D-Dependent AMP Expression in Epidermal Cells

Keratinocytes, mast cells, sebocytes, and other cell types found in the eccrine glands are capable of producing AMPs. Immune cells, such as neutrophils and NK cells, also contribute to the pool of AMPs in the skin [133]. The first skin-derived AMP identified in humans was HBD2, which is triggered by skin inflammation, skin infection, and ultraviolet light B (UV-B) exposure and is highly effective against many gram-negative bacteria [134, 135]. Another well-known cutaneous form of AMP is CAMP, which is comparable to HBD2 and can also be induced by UV-B radiation [136, 137].

In humans, AMP levels increase in both skin infections and injuries [138, 139]. Infectious challenges promote CYP27B1 expression in the cutaneous microenvironment, thereby increasing the local production of 1,25(OH)$_2$D. The presumed mechanism is that skin injury causes an increase in the TLR2 response, leading to elevated vitamin D-induced expression of CAMP and HBD2 [136, 140]. Wang et al. [114] reported that the expression of AMPs in human keratinocytes increased within 24 hours after vitamin D treatment. In the study by Tiosano et al. [15], the exogenous treatment of cells with 25 (OH)D led to an increase in CAMP mRNA expression of more than twofold, whereas β-defensin mRNA expression increased only slightly. Peric et al. [141] demonstrated variable expression of HBD2 mRNA in cultured human epidermal keratinocytes treated with vitamin D analogs.

The application of vitamin D (1,25(OH)$_2$D) in cultured sebocytes, as reported by Lee et al., increases cathelicidin expression [142]. In vivo experiments on the human skin have demonstrated the inductive effects of vitamin D during wound healing. The topical application of calcipotriol enhances the expression of CAMP/LL-37 and helps prevent infection [143]. 1,25 (OH)$_2$D exerts its influence by activating AP-1 and the PPAR-γ pathway, which is induced by p38, leading to the expression of HBD3 and CAMP in human keratinocytes. At the molecular level, the regulatory mechanisms differ slightly between CAMP and HBD2. A 1,25(OH)$_2$D-VDR complex binds to the VDRE in the promoter region of the cathelicidin gene. The expression of HBD2 can also be regulated through the interaction of NF-κB with the 1,25(OH)$_2$D-VDR complex [144]. In addition, 1,25(OH)$_2$D leads to the expression of specific serine proteases, namely, kallikrein-related peptidase (KLK)5 and KLK7. Serine proteases are responsible for the conversion of the hCAP18 precursor into its active antimicrobial form, cathelicidin [144, 145]. Owing to the role of vitamin D as a key player in inflammation, S. Antal et al. suggested that vitamin D might serve as a potent therapeutic agent for the treatment of skin diseases associated with disturbed hCAP18/LL-37 expression [146] (Fig. 5).

Adequate Levels of Vitamin D Maintain AMP Expression

The association between AMP levels and preventive vitamin D supplementation has been examined in several studies. Vargas Buonfiglio et al. [147] reported that vitamin D3 supplementation for 90 days increases the antimicrobial activity of airway surface liquid (ASL) in nonsmoking individuals. This effect is likely mediated by elevated levels of AMPs in the ASL. In contrast, smokers did not exhibit any improvement in ASL antimicrobial activity. The authors speculated that this is partly due to impaired conversion of vitamin D3 to its active form, resulting in decreased expression of LL-37.

Raftery et al. [148] evaluated the relationship between vitamin D and AMP levels in patients with Crohn's disease. They reported that the incidence of Crohn's disease was greater among patients with low levels of vitamin D. Supplementation with vitamin D at a dosage of 2000 IU/day for 3 months was correlated with a significant increase in the level of LL-37 (from 78.4 to 93.6 ng/ml, $p = 0.05$), which subsequently contributed to a decrease in the frequency of disease exacerbation.

Strong innate immune responses, including the synthesis of AMPs, are also crucial for oral health. According to a study by Yoshida et al. [149], LL-37, at a concentration of 10 μmol L^{-1}, binds to heparin within a limited range (2–6 μg mL^{-1}), reducing its cytotoxicity to dental

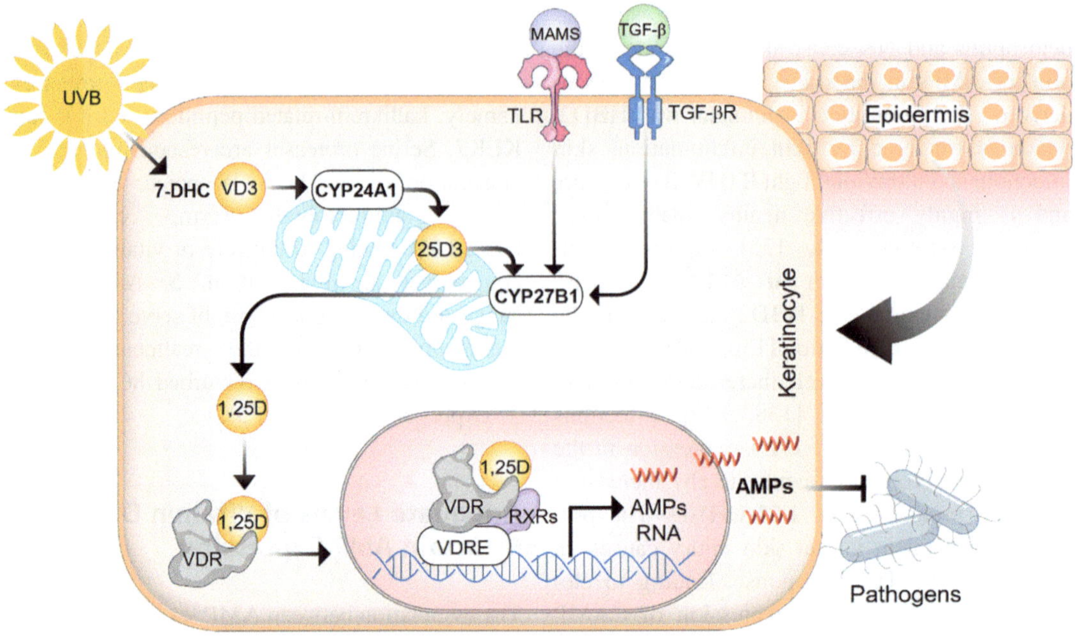

Fig. 5 The effect of vitamin D on AMP expression in keratinocytes. Keratinocytes express VDR and can produce their own 1,25(OH)$_2$D from vitamin D$_3$, which is synthesized from 7-dehydrocholesterol (DHC) under UV-B light. This process is facilitated by CYP27A1, which converts vitamin D$_3$ to 25OHD, and CYP27B1, which converts 25OHD to 1,25(OH)$_2$D. In cases of skin injury or infection, keratinocytes are activated by TGF-β or TLR ligands, which induce the expression of CYP27B1. This leads to the production of 1,25(OH)$_2$D, which stimulates the synthesis of AMPs upon the activation of the VDR

pulp cells while exhibiting strong antimicrobial effects against several bacterial strains, including *Streptococcus mutans*, *Streptococcus sobrinus*, *Streptococcus salivarius*, *Aggregatibacter actinomycetemcomitans*, and *Escherichia coli*. A recent study conducted by Nireeksha et al. [150] revealed that saliva levels of vitamin D significantly decreased with increasing severity of dental caries, indicating that vitamin D is involved in caries prevention. Although the LL-37 level was greater in the caries-free group than in the control group, this difference was not statistically significant.

Low serum vitamin D levels have also been linked to bacterial urinary tract infections. Adequate vitamin D status may enhance the ability of the bladder epithelium to mount a faster and more effective immune response upon microbial invasion [23, 151]. Georgieva et al. reported that vitamin D deficiency is widespread among infants and young children, affecting approximately 21% of the studied population [152]. The study revealed that vitamin D levels in the serum were inversely associated with age and were notably lower in girls. Moreover, there was a positive correlation between vitamin D levels and LL-37, whereas no such correlation was found with β-defensin-2.

Numerous studies have linked low vitamin D levels to the severity of bacterial sepsis [153, 154]. H. Hays et al. reported statistical correlations between seasonal fluctuations in vitamin D status and the incidence of sepsis, which tends to be highest in winter and lowest in fall [155]. A meta-analysis by K. de Haan [156] indicated that vitamin D deficiency increases susceptibility to severe infections and mortality among critically ill patients. Specifically, 25 (OH)D levels less than 50 nmol/L were associated with higher rates of infection (risk ratio (RR) 1.49, 95% confidence interval), sepsis (RR 1.46, 95%), 30-day mortality (RR 1.42, 95%), and in-hospital mortality (RR 1.79, 95%).

Conclusions

Extensive research highlights the crucial role of vitamin D in the innate immune system, particularly through its ability to activate AMP production and initiate appropriate inflammatory responses. The efficacy of these defense mechanisms suggests that vitamin D acts as a potent inducer of immune responses. Vitamin D-induced changes in the expression of genes responsible for the intracellular production of AMPs, coupled with elevated serum levels of 25 (OH)D, promote the production of cathelicidins and defensins, thereby increasing resistance to infectious diseases. To support adjunctive therapy aimed at boosting immunity and preventing bacterial infections, it is essential to maintain sufficient serum levels of 25(OH)D for the optimal function of the antibacterial immune system. A deeper understanding of AMPs as "natural antibiotics" and their role as "immune regulators" opens promising avenues for developing new strategies to combat infectious diseases and address antimicrobial resistance.

Funding This research was funded by the Science Committee of the Ministry of Science and Higher Education of the Republic of Kazakhstan (Grant No. BR24992814).

References

1. Kennel KA, Drake MT, Hurley DL (2010) Vitamin D deficiency in adults: when to test and how to treat. Mayo Clin Proc 85:752–757. quiz 757–758
2. Płudowski P, Karczmarewicz E, Bayer M, Carter G, Chlebna-Sokół D, Czech-Kowalska J, Dębski R, Decsi T, Dobrzańska A, Franek E, Głuszko P, Grant WB, Holick MF, Yankovskaya L, Konstantynowicz J, Książyk JB, Księżopolska-Orłowska K, Lewiński A, Litwin M, Lohner S, Lorenc RS, Lukaszkiewicz J, Marcinowska-Suchowierska E, Milewicz A, Misiorowski W, Nowicki M, Povoroznyuk V, Rozentryt P, Rudenka E, Shoenfeld Y, Socha P, Solnica B, Szalecki M, Tałałaj M, Varbiro S, Żmijewski MA (2013) Practical guidelines for the supplementation of vitamin D and the treatment of deficits in Central Europe – recommended vitamin D intakes in the general population and groups at risk of vitamin D deficiency. Endokrynol Pol 64:319–327
3. Rusińska A, Płudowski P, Walczak M, Borszewska-Kornacka MK, Bossowski A, Chlebna-Sokół D, Czech-Kowalska J, Dobrzańska A, Franek E, Helwich E, Jackowska T, Kalina MA, Konstantynowicz J, Książyk J, Lewiński A, Łukaszkiewicz J, Marcinowska-Suchowierska E, Mazur A, Michałus I, Peregud-Pogorzelski J, Romanowska H, Ruchała M, Socha P, Szalecki M, Wielgoś M, Zwolińska D, Zygmunt A (2018) Vitamin D Supplementation Guidelines for General Population and Groups at Risk of Vitamin D Deficiency in Poland – Recommendations of the Polish Society of Pediatric Endocrinology and Diabetes and the Expert Panel With Participation of National Specialist Consultants and Representatives of Scientific Societies – 2018 Update. Front Endocrinol 9
4. Altieri B, Cavalier E, Bhattoa HP, Perez-Lopez FR, Lopez-Baena MT, Perez-Roncero GR, Chedraui P, Annweiler C, Della Casa S, Zelzer S, Herrmann M, Faggiano A, Colao A, Holick MF (2020) Vitamin D testing: advantages and limits of the current assays. Eur J Clin Nutr 74:231–247
5. Cantorna MT, Snyder L, Lin YD, Yang L (2015) Vitamin D and 1,25(OH)2D regulation of T cells. Nutrients 7:3011–3021
6. Lin R (2016) Crosstalk between Vitamin D metabolism, VDR signalling, and innate immunity. BioMed Res Int 2016:1375858
7. Arboleda JF, Urcuqui-Inchima S (2016) Vitamin D-regulated MicroRNAs: are they protective factors against dengue virus infection? Adv Virol 2016: 1016840
8. Mangin M, Sinha R, Fincher K (2014) Inflammation and vitamin D: the infection connection. Inflamm Res 63:803–819
9. Zdrenghea MT, Makrinioti H, Bagacean C, Bush A, Johnston SL, Stanciu LA (2017) Vitamin D modulation of innate immune responses to respiratory viral infections. Rev Med Virol 27
10. Lu M, Taylor BV, Körner H (2018) Genomic effects of the Vitamin D receptor: potentially the link between Vitamin D, immune cells, and multiple sclerosis. Front Immunol 9:477
11. Agier J, Brzezińska-Błaszczyk E (2016) Cathelicidins and defensins regulate mast cell antimicrobial activity. Postepy Hig Med Dosw (Online) 70: 618–636
12. Skrobot A, Demkow U, Wachowska M (2018) Immunomodulatory role of Vitamin D: a review. Adv Exp Med Biol 1108:13–23
13. Sassi F, Tamone C, D'Amelio P (2018) Vitamin D: nutrient, hormone, and immunomodulator. Nutrients 10
14. Gonzalez-Curiel I, Trujillo V, Montoya-Rosales A, Rincon K, Rivas-Calderon B, deHaro-Acosta J, Marin-Luevano P, Lozano-Lopez D, Enciso-Moreno JA, Rivas-Santiago B (2014) 1,25-dihydroxyvitamin D3 induces LL-37 and HBD-2 production in keratinocytes from diabetic foot ulcers promoting

wound healing: an in vitro model. PLoS One 9: e111355
15. Tiosano D, Wildbaum G, Gepstein V, Verbitsky O, Weisman Y, Karin N, Eztioni A (2013) The role of vitamin D receptor in innate and adaptive immunity: a study in hereditary vitamin D-resistant rickets patients. J Clin Endocrinol Metab 98:1685–1693
16. Zhang LJ, Gallo RL (2016) Antimicrobial peptides. Curr Biol 26:R14–R19
17. Jeng L, Yamshchikov AV, Judd SE, Blumberg HM, Martin GS, Ziegler TR, Tangpricha V (2009) Alterations in vitamin D status and anti-microbial peptide levels in patients in the intensive care unit with sepsis. J Transl Med 7:28
18. Kempker JA, Tangpricha V, Ziegler TR, Martin GS (2012) Vitamin D in sepsis: from basic science to clinical impact. Crit Care 16:316
19. Quraishi SA, De Pascale G, Needleman JS, Nakazawa H, Kaneki M, Bajwa EK, Camargo CA Jr, Bhan I (2015) Effect of cholecalciferol supplementation on Vitamin D status and cathelicidin levels in sepsis: a randomized, placebo-controlled trial. Crit Care Med 43:1928–1937
20. Çekmez Y, Öcal D, Korkmaz V, Sertoğlu E, Uçar Y, Küçüközkan T (2015) The diagnostic role of Vitamin D and cathelicidin levels in pelvic inflammatory disease. Clin Lab 61:1871–1875
21. Yim S, Dhawan P, Ragunath C, Christakos S, Diamond G (2007) Induction of cathelicidin in normal and CF bronchial epithelial cells by 1,25-dihydroxyvitamin D(3). J Cyst Fibros 6:403–410
22. Georgieva V, Kamolvit W, Herthelius M, Lüthje P, Brauner A, Chromek M (2019) Association between vitamin D, antimicrobial peptides and urinary tract infection in infants and young children. Acta Paediatr 108:551–556
23. Hertting O, Holm Å, Lüthje P, Brauner H, Dyrdak R, Jonasson AF, Wiklund P, Chromek M, Brauner A (2010) Vitamin D induction of the human antimicrobial peptide cathelicidin in the urinary bladder. PLoS One 5:e15580
24. Nielsen OH, Hansen TI, Gubatan JM, Jensen KB, Rejnmark L (2019) Managing vitamin D deficiency in inflammatory bowel disease. Frontline Gastroenterol 10:394–400
25. Tabatabaeizadeh SA, Tafazoli N, Ferns GA, Avan A, Ghayour-Mobarhan M (2018) Vitamin D, the gut microbiome and inflammatory bowel disease. J Res Med Sci 23:75
26. Vinh Quôc Luong K, Thi Hoàng Nguyên L (2012) Role of the Vitamin D in leprosy. Am J Med Sci 343: 471–482
27. Matzner M, Al Samie AR, Winkler HM, Nemeth J, Grasnek A, Indra A, Bieglmayer C, Winkler S (2011) Low serum levels of cathelicidin LL-37 in leprosy. Acta Trop 117:56–59
28. Khammissa RAG, Ballyram R, Jadwat Y, Fourie J, Lemmer J, Feller L (2018) Vitamin D deficiency as it relates to oral immunity and chronic periodontitis. Int J Dent 2018:7315797
29. Jagelavičienė E, Vaitkevičienė I, Šilingaitė D, Šinkūnaitė E, Daugėlaitė G (2018) The relationship between Vitamin D and periodontal pathology, Medicina (Kaunas). 54
30. Duan X, Guan Y, Li Y, Chen S, Li S, Chen L (2015) Vitamin D potentiates the inhibitory effect of MicroRNA-130a in hepatitis C Virus replication independent of Type I interferon signaling pathway. Mediat Inflamm 2015:508989
31. Zasloff M (2019) Antimicrobial peptides of multicellular organisms: my perspective. Adv Exp Med Biol 1117:3–6
32. Lorenzon EN, Piccoli JP, Santos-Filho NA, Cilli EM (2019) Dimerization of antimicrobial peptides: a promising strategy to enhance antimicrobial peptide activity. Protein Pept Lett 26:98–107
33. Liu S, Fan L, Sun J, Lao X, Zheng H (2017) Computational resources and tools for antimicrobial peptides. J Pept Sci 23:4–12
34. Barzyk W, Rogalska E, Więcław-Czapla K (2013) Penetration of milk-derived antimicrobial peptides into phospholipid monolayers as model biomembranes. Biochem Res Int 2013:914540
35. Lei J, Sun L, Huang S, Zhu C, Li P, He J, Mackey V, Coy DH, He Q (2019) The antimicrobial peptides and their potential clinical applications. Am J Transl Res 11:3919–3931
36. Bhattacharjya S, Zhang Z, Ramamoorthy A (2024) LL-37: structures, antimicrobial activity, and influence on amyloid-related diseases. Biomolecules 14(3):320. https://doi.org/10.3390/biom14030320
37. Agier J, Efenberger M, Brzezińska-Błaszczyk E (2015) Cathelicidin impact on inflammatory cells. Cent Eur J Immunol 40:225–235
38. Méndez-Samperio P (2010) The human cathelicidin hCAP18/LL-37: a multifunctional peptide involved in mycobacterial infections. Peptides 31:1791–1798
39. Zhao C, Nguyen T, Liu L, Sacco RE, Brogden KA, Lehrer RI (2001) Gallinacin-3, an inducible epithelial beta-defensin in the chicken. Infect Immun 69:2684–2691
40. Meisch JP, Vogel RM, Schlatzer DM, Li X, Chance MR, Levine AD (2013) Human β-defensin 3 induces STAT1 phosphorylation, tyrosine phosphatase activity, and cytokine synthesis in T cells. J Leukoc Biol 94:459–471
41. Lee JY, Suh JS, Kim JM, Kim JH, Park HJ, Park YJ, Chung CP (2015) Identification of a cell-penetrating peptide domain from human beta-defensin 3 and characterization of its anti-inflammatory activity. Int J Nanomedicine 10:5423–5434
42. Zhao BC, Lin HC, Yang D, Ye X, Li ZG (2015) Disulfide bridges in defensins. Curr Top Med Chem 16:206–219
43. Niyonsaba F, Ushio H, Hara M, Yokoi H, Tominaga M, Takamori K, Kajiwara N, Saito H, Nagaoka I, Ogawa H, Okumura K (2010)

Antimicrobial peptides human beta-defensins and cathelicidin LL-37 induce the secretion of a pruritogenic cytokine IL-31 by human mast cells. J Immunol 184:3526–3534
44. Frohm Nilsson M, Sandstedt B, Sørensen O, Weber G, Borregaard N, Ståhle-Bäckdahl M (1999) The human cationic antimicrobial protein (hCAP18), a peptide antibiotic, is widely expressed in human squamous epithelia and colocalizes with interleukin-6. Infect Immun 67:2561–2566
45. Kahlenberg JM, Kaplan MJ (2013) Little peptide, big effects: the role of LL-37 in inflammation and autoimmune disease. J Immunol 191:4895–4901
46. Findlay F, Proudfoot L, Stevens C, Barlow PG (2016) Cationic host defense peptides; novel antimicrobial therapeutics against Category A pathogens and emerging infections. Pathog Glob Health 110: 137–147
47. Faurschou M, Kamp S, Cowland JB, Udby L, Johnsen AH, Calafat J, Winther H, Borregaard N (2005) Prodefensins are matrix proteins of specific granules in human neutrophils. J Leukoc Biol 78: 785–793
48. Li X, Duan D, Yang J, Wang P, Han B, Zhao L, Jepsen S, Dommisch H, Winter J, Xu Y (2016) The expression of human β-defensins (hBD-1, hBD-2, hBD-3, hBD-4) in gingival epithelia. Arch Oral Biol 66:15–21
49. Vandamme D, Landuyt B, Luyten W, Schoofs L (2012) A comprehensive summary of LL-37, the factotum human cathelicidin peptide. Cell Immunol 280:22–35
50. Bikle DD (2022) Vitamin D regulation of immune function. Curr Osteoporos Rep 20:186–193
51. Li X, Zuo S, Wang B, Zhang K, Wang Y (2022) Antimicrobial mechanisms and clinical application prospects of antimicrobial peptides. In Molecules 27(9):2675. https://doi.org/10.3390/molecules27092675
52. Malanovic N, Lohner K (2016) Gram-positive bacterial cell envelopes: the impact on the activity of antimicrobial peptides. Biochim Biophys Acta 1858:936–946
53. Fuertes G, Giménez D, Esteban-Martín S, Sánchez-Muñoz OL, Salgado J (2011) A lipocentric view of peptide-induced pores. Eur Biophys J 40:399–415
54. Rashid R, Veleba M, Kline KA (2016) Focal targeting of the bacterial envelope by antimicrobial peptides. Front Cell Dev Biol 4:55–55
55. Wimley WC (2010) Describing the mechanism of antimicrobial peptide action with the interfacial activity model. ACS Chem Biol 5:905–917
56. Chang WK, Wimley WC, Searson PC, Hristova K, Merzlyakov M (2008) Characterization of antimicrobial peptide activity by electrochemical impedance spectroscopy. Biochim Biophys Acta 1778:2430–2436
57. Zhang M, Zhao J, Zheng J (2014) Molecular understanding of a potential functional link between antimicrobial and amyloid peptides. Soft Matter 10: 7425–7451
58. López-Meza JE, Aguilar A, Loeza-Lara PD (2011) Antimicrobial peptides: diversity and perspectives for their biomedical application. InTech https://doi.org/10.5772/13058
59. Tang M, Hong M (2009) Structure and mechanism of beta-hairpin antimicrobial peptides in lipid bilayers from solid-state NMR spectroscopy. Mol Biosyst 5: 317–322
60. Bechinger B, Lohner K (2006) Detergent-like actions of linear amphipathic cationic antimicrobial peptides. Biochim Biophys Acta 1758:1529–1539
61. Luo Y, Song Y (2021) Mechanism of antimicrobial peptides: antimicrobial, anti-inflammatory and antibiofilm activities. Int J Mol Sci 22
62. Sim S, Wang P, Beyer BN, Cutrona KJ, Radhakrishnan ML, Elmore DE (2017) Investigating the nucleic acid interactions of histone-derived antimicrobial peptides. FEBS Lett 591:706–717
63. Marchand C, Krajewski K, Lee HF, Antony S, Johnson AA, Amin R, Roller P, Kvaratskhelia M, Pommier Y (2006) Covalent binding of the natural antimicrobial peptide indolicidin to DNA abasic sites. Nucleic Acids Res 34:5157–5165
64. Sugiarto H, Yu P-L (2007) Mechanisms of action of ostrich β-defensins against Escherichia coli. FEMS Microbiology Letters 270:195–200
65. Xhindoli D, Pacor S, Benincasa M, Scocchi M, Gennaro R, Tossi A (2016) The human cathelicidin LL-37 – a pore-forming antibacterial peptide and host-cell modulator. Biochimica et Biophysica Acta (BBA) Biomembranes 1858:546–566
66. Wang C, Zhao G, Wang S, Chen Y, Gong Y, Chen S, Xu Y, Hu M, Wang X, Zeng H, Wang A, Liu D, Su Y, Cheng T, Chen F, Wang J (2018) A simplified derivative of human Defensin 5 with potent and efficient activity against multidrug-resistant Acinetobacter baumannii. Antimicrob Agents Chemother 62. https://doi.org/10.1128/aac.01504-01517
67. Routsias JG, Karagounis P, Parvulesku G, Legakis NJ, Tsakris A (2010) In vitro bactericidal activity of human β-defensin 2 against nosocomial strains. Peptides 31:1654–1660
68. Maisetta G, Batoni G, Esin S, Florio W, Bottai D, Favilli F, Campa M (2006) In vitro bactericidal activity of human beta-defensin 3 against multidrug-resistant nosocomial strains. Antimicrob Agents Chemother 50:806–809
69. Supp DM, Gardner J, Klingenberg JM, Neely AN (2009) Antibiotic resistance in clinical isolates of Acinetobacter baumannii, Pseudomonas aeruginosa, and Staphylococcus aureus does not impact sensitivity to human beta defensin 4. Burns 35:949–955
70. Maisetta G, Batoni G, Esin S, Raco G, Bottai D, Favilli F, Florio W, Campa M (2005) Susceptibility of Streptococcus mutans and Actinobacillus actinomycetemcomitans to bactericidal activity of human beta-defensin 3 in biological fluids. Antimicrob Agents Chemother 49:1245–1248
71. Ericksen B, Wu Z, Lu W, Lehrer RI (2005) Antibacterial activity and specificity of the six

human {alpha}-defensins. Antimicrob Agents Chemother 49:269–275
72. Mathew B, Nagaraj R (2015) Antimicrobial activity of human α-defensin 5 and its linear analogs: N-terminal fatty acylation results in enhanced antimicrobial activity of the linear analogs. Peptides 71: 128–140
73. So YJ, Park O-J, Kwon Y, Im J, Lee D, Yun SH, Cho K, Yun C-H, Han SH (2024) Bacillus subtilis induces human beta Defensin-2 through its lipoproteins in human intestinal epithelial cells. Probiot Antimicrob Prot
74. Schroeder BO, Wu Z, Nuding S, Groscurth S, Marcinowski M, Beisner J, Buchner J, Schaller M, Stange EF, Wehkamp J (2011) Reduction of disulphide bonds unmasks potent antimicrobial activity of human β-defensin 1. Nature 469:419–423
75. Schroeder BO, Ehmann D, Precht JC, Castillo PA, Küchler R, Berger J, Schaller M, Stange EF, Wehkamp J (2015) Paneth cell α-defensin 6 (HD-6) is an antimicrobial peptide. Mucosal Immunol 8:661–671
76. Furci L, Baldan R, Bianchini V, Trovato A, Ossi C, Cichero P, Cirillo DM (2015) New role for human α-defensin 5 in the fight against hypervirulent Clostridium difficile strains. Infect Immun 83:986–995
77. Supandy A, Van Tyne D (2025) AMPing up the pressure: cell envelope signaling protects enterococcus faecalis from antimicrobial peptides. J Infect Dis 231:287–290
78. Kandaswamy K, Liew TH, Wang CY, Huston-Warren E, Meyer-Hoffert U, Hultenby K, Schröder JM, Caparon MG, Normark S, Henriques-Normark B, Hultgren SJ, Kline KA (2013) Focal targeting by human β-defensin 2 disrupts localized virulence factor assembly sites in Enterococcus faecalis. Proc Natl Acad Sci 110:20230
79. Ahn KB, Kim AR, Kum KY, Yun CH, Han SH (2017) The synthetic human beta-defensin-3 C15 peptide exhibits antimicrobial activity against Streptococcus mutans, both alone and in combination with dental disinfectants. J Microbiol 55:830–836
80. Turner J, Cho Y, Dinh NN, Waring AJ, Lehrer RI (1998) Activities of LL-37, a cathelin-associated antimicrobial peptide of human neutrophils. Antimicrob Agents Chemother 42:2206–2214
81. Morroni G, Di SL, Oriana S, Lucia B, Wojciech K, Elzbieta K, Marina M, Andrea B, Eleonora G, Patrizia B, Andrea G, Cirioni O (2021) Synergistic effect of antimicrobial peptide LL-37 and colistin combination against multidrug-resistant Escherichia coli isolates. Fut Microbiol 16:221–227
82. Ganz T, Selsted ME, Szklarek D, Harwig SS, Daher K, Bainton DF, Lehrer RI (1985) Defensins. Natural peptide antibiotics of human neutrophils. J Clin Invest 76:1427–1435
83. Pazgier M, Lubkowski J (2006) Expression and purification of recombinant human α-defensins in Escherichia coli. Prot Express Purif 49:1–8
84. Varney KM, Bonvin AM, Pazgier M, Malin J, Yu W, Ateh E, Oashi T, Lu W, Huang J, Diepeveen-de Buin M, Bryant J, Breukink E, Mackerell AD Jr, de Leeuw EP (2013) Turning defense into offense: defensin mimetics as novel antibiotics targeting lipid II. PLoS Pathog 9:e1003732
85. Maiti S, Patro S, Purohit S, Jain S, Senapati S, Dey N (2014) Effective control of Salmonella infections by employing combinations of recombinant antimicrobial human β-defensins hBD-1 and hBD-2. Antimicrob Agents Chemother 58:6896–6903
86. Wilde CG, Griffith JE, Marra MN, Snable JL, Scott RW (1989) Purification and characterization of human neutrophil peptide 4, a novel member of the defensin family. J Biol Chem 264:11200–11203
87. Patro S, Maiti S, Panda SK, Dey N (2015) Utilization of plant-derived recombinant human β-defensins (hBD-1 and hBD-2) for averting salmonellosis. Transgenic Res 24:353–364
88. Sahl HG, Pag U, Bonness S, Wagner S, Antcheva N, Tossi A (2005) Mammalian defensins: structures and mechanism of antibiotic activity. J Leukoc Biol 77: 466–475
89. Schröder JM, Harder J (1999) Human beta-defensin-2. Int J Biochem Cell Biol 31:645–651
90. McGlasson SL, Semple F, MacPherson H, Gray M, Davidson DJ, Dorin JR (2017) Human β-defensin 3 increases the TLR9-dependent response to bacterial DNA. Eur J Immunol 47:658–664
91. Rivera KG, Tanaka KJ, Buechel ER, Origel O Jr, Harrison A, Mason KM, Pinkett HW (2024) Antimicrobial peptide recognition motif of the substrate binding protein SapA from nontypeable Haemophilus influenzae. Biochemistry 63:294–311
92. Lee HY, Andalibi A, Webster P, Moon SK, Teufert K, Kang SH, Li JD, Nagura M, Ganz T, Lim DJ (2004) Antimicrobial activity of innate immune molecules against Streptococcus pneumoniae, Moraxella catarrhalis and nontypeable Haemophilus influenzae. BMC Infect Dis 4:12
93. Milad N, Pineault M, Bouffard G, Maranda-Robitaille M, Lechasseur A, Beaulieu M-J, Aubin S, Jensen BAH, Morissette MC (2022) Recombinant human β-defensin 2 delivery improves smoking-induced lung neutrophilia and bacterial exacerbation. Am J Physiol Lung Cell Mol Physiol 323:L37–L47
94. Escobar-Salom M, Barceló Isabel M, Rojo-Molinero-E, Jordana-Lluch E, Cabot G, Oliver A, Juan C (2024) In vitro activity of human defensins HNP-1 and hBD-3 against multidrug-resistant ESKAPE Gram-negatives of clinical origin and selected peptidoglycan recycling-defective mutants. Microbiol Spectr 12:e00358–e00324
95. Rahman S, Das AK (2022) A subtractive proteomics and immunoinformatics approach towards designing a potential multi-epitope vaccine against pathogenic Listeria monocytogenes. Microb Pathog 172:105782
96. Ogata K, Linzer BA, Zuberi RI, Ganz T, Lehrer RI, Catanzaro A (1992) Activity of defensins from human neutrophilic granulocytes against Mycobacterium avium-Mycobacterium intracellulare. Infect Immun 60:4720–4725

97. Li X, Xiaoming W, Jiajing D, Xiangzhen B, Chao P, Ximeng D, Fu C (2024) Applications of β-defensins against infectious pathogenic microorganisms. Exp Rev Anti-infective Ther 22:501–510
98. Neidig A, Strempel N, Waeber NB, Nizer WSDC, Overhage J (2023) Knock-out of multidrug efflux pump MexXY-OprM results in increased susceptibility to antimicrobial peptides in Pseudomonas aeruginosa. Microb Immun 67:422–427
99. Fusco A, Savio V, Stelitano D, Baroni A, Donnarumma G (2021) The intestinal biofilm of Pseudomonas aeruginosa and Staphylococcus aureus is inhibited by antimicrobial peptides HBD-2 and HBD-3. Appl Sci 11
100. Heffernan Linda M, Lawrence Anna-Lisa E, Marcotte Haley A, Sharma A, Jenkins Aria X, Iguwe D, Rood J, Herke Scott W, O'Riordan Mary X, Abuaita Basel H (2024) Heterogeneity of Salmonella enterica lipopolysaccharide counteracts macrophage and antimicrobial peptide defenses. Infect Immun 92:e00251–e00224
101. Fusco A, Savio V, Cammarota M, Alfano A, Schiraldi C, Donnarumma G (2017) Beta-Defensin-2 and Beta-Defensin-3 reduce intestinal damage caused by Salmonella typhimurium modulating the expression of cytokines and enhancing the probiotic activity of Enterococcus faecium. J Immunol Res 2017:6976935
102. Chu H, Pazgier M, Jung G, Nuccio S-P, Castillo Patricia A, de Jong Maarten F, Winter Maria G, Winter Sebastian E, Wehkamp J, Shen B, Salzman Nita H, Underwood Mark A, Tsolis Renee M, Young Glenn M, Lu W, Lehrer Robert I, Bäumler Andreas J, Bevins Charles L (2012) Human α-Defensin 6 promotes mucosal innate immunity through self-assembled peptide nanonets. Science 337:477–481
103. Ridyard KE, Overhage J (2021) The potential of human peptide LL-37 as an antimicrobial and anti-biofilm agent. Antibiotics 10
104. Kraemer BF, Campbell RA, Schwertz H, Cody MJ, Franks Z, Tolley ND, Kahr WH, Lindemann S, Seizer P, Yost CC, Zimmerman GA, Weyrich AS (2011) Novel anti-bacterial activities of β-defensin 1 in human platelets: suppression of pathogen growth and signaling of neutrophil extracellular trap formation. PLoS Pathog 7:e1002355
105. Sass V, Schneider T, Wilmes M, Körner C, Tossi A, Novikova N, Shamova O, Sahl HG (2010) Human beta-defensin 3 inhibits cell wall biosynthesis in Staphylococci. Infect Immun 78:2793–2800
106. Mori T, Yoshida M, Hazekawa M, Ishibashi D, Hatanaka Y, Nagao T, Kakehashi R, Kojima H, Uno R, Ozeki M, Kawasaki I, Yamashita T, Nishikawa J, Uchida T (2021) Antimicrobial activities of LL-37 fragment mutant-poly (Lactic-Co-Glycolic) acid conjugate against Staphylococcus aureus, Escherichia coli, and Candida albicans. Int J Mol Sci 22
107. Ying X, Xue G, Sun P, Gan Z, Fan Z, Liu B, Han Y, Yang J, Zhang J, Lu A (2023) Antimicrobial peptides targeting Streptococcus mutans: current research on design, screening and efficacy. Curr Microbiol 81:18
108. Lin L, Wen S-H, Guo S-Z, Su X-Y, Wu H-J, Chong L, Zhang H-L, Zhang W-X, Li C-C (2014) Role of SIRT1 in Streptococcus pneumoniae-induced human β-defensin-2 and interleukin-8 expression in A549 cell. Mol Cell Biochem 394:199–208
109. Buonfiglio LGV, Cano M, Pezzulo AA, Calderon OGV, Zabner J, Gerke AK, Comellas AP (2017) Effect of vitamin D 3 on the antimicrobial activity of human airway surface liquid: preliminary results of a randomised placebo-controlled double-blind trial. BMJ Open Respir Res 4:e000211
110. Edfeldt K, Liu PT, Chun R, Fabri M, Schenk M, Wheelwright M, Keegan C, Krutzik SR, Adams JS, Hewison M (2010) T-cell cytokines differentially control human monocyte antimicrobial responses by regulating vitamin D metabolism. Proc Natl Acad Sci 107:22593–22598
111. Liu PT, Schenk M, Walker VP, Dempsey PW, Kanchanapoomi M, Wheelwright M, Vazirnia A, Zhang X, Steinmeyer A, Zügel U (2009) Convergence of IL-1β and VDR activation pathways in human TLR2/1-induced antimicrobial responses. PLoS One 4:e5810
112. Schauber J, Dorschner RA, Coda AB, Büchau AS, Liu PT, Kiken D, Helfrich YR, Kang S, Elalieh HZ, Steinmeyer A (2007) Injury enhances TLR2 function and antimicrobial peptide expression through a vitamin D–dependent mechanism. J Clin Invest 117: 803–811
113. Gombart AF, Borregaard N, Koeffler HP (2005) Human cathelicidin antimicrobial peptide (CAMP) gene is a direct target of the vitamin D receptor and is strongly up-regulated in myeloid cells by 1, 25-dihydroxyvitamin D3. FASEB J 19:1067–1077
114. Wang TT, Nestel FP, Bourdeau V, Nagai Y, Wang Q, Liao J, Tavera-Mendoza L, Lin R, Hanrahan JW, Mader S, White JH (2004) Cutting edge: 1,25-dihydroxyvitamin D3 is a direct inducer of antimicrobial peptide gene expression. J Immunol 173:2909–2912
115. White JH (2022) Emerging roles of Vitamin D-induced antimicrobial peptides in antiviral innate immunity. Nutrients:14
116. Dimitrov V, White JH (2016) Species-specific regulation of innate immunity by Vitamin D signaling. J Steroid Biochem Mol Biol 164:246–253
117. Takeda K, Akira S (2005) Toll-like receptors in innate immunity. Int Immunol 17:1–14
118. Kawai T, Akira S (2010) The role of pattern-recognition receptors in innate immunity: update on Toll-like receptors. Nat Immunol 11:373–384
119. Liu PT, Stenger S, Li H, Wenzel L, Tan BH, Krutzik SR, Ochoa MT, Schauber J, Wu K, Meinken C, Kamen DL, Wagner M, Bals R, Steinmeyer A, Zügel U, Gallo RL, Eisenberg D, Hewison M, Hollis

BW, Adams JS, Bloom BR, Modlin RL (2006) Toll-like receptor triggering of a vitamin D-mediated human antimicrobial response. Science 311:1770–1773
120. Rode AKO, Kongsbak M, Hansen MM, Lopez DV, Levring TB, Woetmann A, Ødum N, Bonefeld CM, Geisler C (2017) Vitamin D counteracts mycobacterium tuberculosis-induced cathelicidin downregulation in dendritic cells and allows Th1 differentiation and IFNγ secretion. Front Immunol 8:656
121. Ashenafi S, Mazurek J, Rehn A, Lemma B, Aderaye G, Bekele A, Assefa G, Chanyalew M, Aseffa A, Andersson J, Bergman P, Brighenti S (2018) Vitamin D_3 status and the association with human cathelicidin expression in patients with different clinical forms of active tuberculosis. Nutrients 10
122. Iftikhar R, Kamran SM, Qadir A, Haider E, Bin Usman H (2013) Vitamin D deficiency in patients with tuberculosis. J Coll Physicians Surg Pak 23: 780–783
123. Adams JS, Hewison M (2008) Unexpected actions of vitamin D: new perspectives on the regulation of innate and adaptive immunity. Nat Clin Pract Endocrinol Metabol 4:80–90
124. Zhao S, He Y, Pan M, Chen B, Zhang S, Zhang Y, Zhu Y (2022) Expression and significance of serum vitamin D and LL-37 levels in infants with bacterial pneumonia. Front Pediatr 10
125. Aidoukovitch A, Bankell E, Svensson D, Nilsson B-O (2024) Vitamin D triggers hCAP18/LL-37 production: implications for LL-37-induced human osteoblast cytotoxicity. Biochem Biophys Res Commun 712–713:149962
126. Sakaki T, Kagawa N, Yamamoto K, Inouye K (2005) Metabolism of vitamin D3 by cytochromes P450. Front Biosci 10:119–134
127. Hamza FN, Daher S, Fakhoury HMA, Grant WB, Kvietys PR, Al-Kattan K (2023) Immunomodulatory properties of Vitamin D in the intestinal and respiratory systems. Nutrients 15
128. Wang T-T, Nestel FP, Bourdeau VR, Nagai Y, Wang Q, Liao J, Tavera-Mendoza L, Lin R, Hanrahan JW, Mader S, White JH (2004) Cutting Edge: 1,25-dihydroxyvitamin D3 is a direct inducer of antimicrobial peptide gene expression1. J Immunol 173:2909–2912
129. Wang T-T, Dabbas B, Laperriere D, Bitton AJ, Soualhine H, Tavera-Mendoza LE, Dionne S, Servant MJ, Bitton A, Seidman EG (2010) Direct and indirect induction by 1, 25-dihydroxyvitamin D3 of the NOD2/CARD15-defensin β2 innate immune pathway defective in Crohn disease. J Biol Chem 285:2227–2231
130. Zhang Y-G, Shaoping W, Sun J (2013) Vitamin D, vitamin D receptor and tissue barriers. Tissue Barriers 1:e23118
131. Zhao H, Zhang H, Wu H, Li H, Liu L, Guo J, Li C, Shih DQ, Zhang X (2012) Protective role of 1,25 (OH)2vitamin D3 in the mucosal injury and epithelial barrier disruption in DSS-induced acute colitis in mice. BMC Gastroenterol 12:57
132. Assa A, Vong L, Pinnell LJ, Avitzur N, Johnson-Henry KC, Sherman PM (2014) Vitamin D Deficiency promotes epithelial barrier dysfunction and intestinal inflammation. J Infect Dis 210:1296–1305
133. Dorschner RA, Pestonjamasp VK, Tamakuwala S, Ohtake T, Rudisill J, Nizet V, Agerberth B, Gudmundsson GH, Gallo RL (2001) Cutaneous injury induces the release of cathelicidin antimicrobial peptides active against group A Streptococcus. J Invest Dermatol 117:91–97
134. Zaiou M, Gallo RL (2002) Cathelicidins, essential gene-encoded mammalian antibiotics. J Mol Med (Berl) 80:549–561
135. Schauber J, Gallo RL (2007) Expanding the roles of antimicrobial peptides in skin: alarming and arming keratinocytes. J Invest Dermatol 127:510–512
136. Schauber J, Dorschner RA, Coda AB, Büchau AS, Liu PT, Kiken D, Helfrich YR, Kang S, Elalieh HZ, Steinmeyer A, Zügel U, Bikle DD, Modlin RL, Gallo RL (2007) Injury enhances TLR2 function and antimicrobial peptide expression through a vitamin D–dependent mechanism. J Clin Invest 117:803–811
137. Yamasaki K, Schauber J, Coda A, Lin H, Dorschner RA, Schechter NM, Bonnart C, Descargues P, Hovnanian A, Gallo RL (2006) Kallikrein-mediated proteolysis regulates the antimicrobial effects of cathelicidins in skin. FASEB J 20:2068–2080
138. Weber G, Heilborn JD, Chamorro Jimenez CI, Hammarsjo A, Törmä H, Stahle M (2005) Vitamin D induces the antimicrobial protein hCAP18 in human skin. J Invest Dermatol 124:1080–1082
139. Umar M, Sastry KS, Al Ali F, Al-Khulaifi M, Wang E, Chouchane AI (2018) Vitamin D and the pathophysiology of inflammatory skin diseases. Skin Pharmacol Physiol 31:74–86
140. Samanta S (2021) Vitamin D and immunomodulation in the skin: a useful affirmative nexus. Explor Immunol:90–111
141. Peric M, Koglin S, Dombrowski Y, Groß K, Bradac E, Büchau A, Steinmeyer A, Zügel U, Ruzicka T, Schauber J (2009) Vitamin D analogs differentially control antimicrobial peptide/"alarmin" expression in psoriasis. PLoS One 4:e6340
142. Lee WJ, Cha HW, Sohn MY, Lee SJ, Kim DW (2012) Vitamin D increases expression of cathelicidin in cultured sebocytes. Arch Dermatol Res 304:627–632
143. Dai X, Sayama K, Tohyama M, Shirakata Y, Hanakawa Y, Tokumaru S, Yang L, Hirakawa S, Hashimoto K (2010) PPARγ mediates innate immunity by regulating the 1α,25-dihydroxyvitamin D3 induced hBD-3 and cathelicidin in human keratinocytes. J Dermatol Sci 60:179–186
144. Liu PT, Schenk M, Walker VP, Dempsey PW, Kanchanapoomi M, Wheelwright M, Vazirnia A, Zhang X, Steinmeyer A, Zügel U, Hollis BW,

Cheng G, Modlin RL (2009) Convergence of IL-1beta and VDR activation pathways in human TLR2/1-induced antimicrobial responses. PLoS One 4:e5810
145. Morizane S, Yamasaki K, Kabigting FD, Gallo RL (2010) Kallikrein expression and cathelicidin processing are independently controlled in keratinocytes by calcium, vitamin D(3), and retinoic acid. J Invest Dermatol 130:1297–1306
146. Antal AS, Dombrowski Y, Koglin S, Ruzicka T, Schauber J (2011) Impact of vitamin D3 on cutaneous immunity and antimicrobial peptide expression. Dermatoendocrinol 3:18–22
147. Luis GVB, Marlene C, Alejandro AP, Oriana GVC, Joseph Z, Alicia KG, Alejandro PC (2017) Effect of vitamin D 3 on the antimicrobial activity of human airway surface liquid: preliminary results of a randomised placebo-controlled double-blind trial. BMJ Open Respir Res 4:e000211
148. Raftery T, Martineau AR, Greiller CL, Ghosh S, McNamara D, Bennett K, Meddings J, O'Sullivan M (2015) Effects of vitamin D supplementation on intestinal permeability, cathelicidin and disease markers in Crohn's disease: results from a randomised double-blind placebo-controlled study. United Eur Gastroenterol J 3:294–302
149. Yoshida K, Suzuki S, Kawada-Matsuo M, Nakanishi J, Hirata-Tsuchiya S, Komatsuzawa H, Yamada S, Shiba H (2019) Heparin–LL37 complexes are less cytotoxic for human dental pulp cells and have undiminished antimicrobial and LPS-neutralizing abilities. Int Endodont J 52:1327–1343
150. Nireeksha, Hegde MN, Kumari NS (2024) Potential role of salivary vitamin D antimicrobial peptide LL-37 and interleukins in severity of dental caries: an exvivo study. BMC Oral Health 24:79
151. Deng QF, Chu H, Wen Z, Cao YS (2019) Vitamin D and urinary tract infection: a systematic review and meta-analysis. Ann Clin Lab Sci 49:134–142
152. Georgieva V, Kamolvit W, Herthelius M, Lüthje P, Brauner A, Chromek M (2019) Association between vitamin D, antimicrobial peptides and urinary tract infection in infants and young children. Acta Paediatrica 108:551–556
153. Danai PA, Sinha S, Moss M, Haber MJ, Martin GS (2007) Seasonal variation in the epidemiology of sepsis. Crit Care Med 35:410–415
154. Kempker JA, Han JE, Tangpricha V, Ziegler TR, Martin GS (2012) Vitamin D and sepsis. Dermato-Endocrinology 4:101–108
155. Hays H, Flores LE, Kothari V, Bilek L, Geske J, Skinner A (2022) Vitamin D status and seasonal variation: a retrospective single institution database study of patients pursuing metabolic/bariatric surgery. Clin Nutr Open Sci 41:1–9
156. de Haan K, Groeneveld AB, de Geus HR, Egal M, Struijs A (2014) Vitamin D deficiency as a risk factor for infection, sepsis and mortality in the critically ill: systematic review and meta-analysis. Crit Care 18:660

Vitamin D and Autoimmunity

Fabrícia Lima Fontes-Dantas, Rafaella de Carvalho Cardoso, Vinícius Domingues, and Gabriel Gripp Fernandes

Abstract

Vitamin D is increasingly recognized as a crucial factor in immune regulation, particularly in the context of autoimmune diseases. In addition to its well-established role in bone metabolism and calcium homeostasis, recent research highlights its significant influence on both adaptive and innate immune responses. In conditions such as multiple sclerosis, type 1 diabetes, and systemic lupus erythematosus, vitamin D enhances the function of regulatory T cells and suppresses pro-inflammatory cytokines, contributing to immune balance. Genetic factors also play a vital role in susceptibility to autoimmune diseases. Studies have investigated the interaction between vitamin D signaling pathways and genetic variants in immune-related genes, including the vitamin D receptor (*VDR*), vitamin D binding protein (*VDBP*), 25-hydroxylase (*CYP2R1*), and 1-α-hydroxylase (*CYP27B1*) genes. Polymorphisms in these genes can alter vitamin D signaling, potentially affecting disease susceptibility and severity. This chapter explains the complex relationships among serum vitamin D levels, genetic variations, and immune regulation in autoimmune disorders. These findings emphasize the importance of further research to clarify the underlying molecular mechanisms and develop tailored therapeutic strategies. A deeper understanding of these interactions could pave the way for innovative preventive and therapeutic approaches for individuals with a genetic predisposition to autoimmune conditions.

Keywords

Vitamin D · Autoimmune diseases · Immune regulation · Genetic variants

Vitamin D and Signaling: An Overview

Vitamin D, which is traditionally recognized for its essential role in regulating calcium metabolism and maintaining bone health, has garnered significant attention in the field of immunology over the past few decades [40, 42]. Insufficient levels of

F. L. Fontes-Dantas (✉) · R. de Carvalho Cardoso
Neuropharmacogenetics Laboratory, Department of Pharmacology and Psychobiology, Roberto Alcântara Gomes Institute Biology (IBRAG), Rio de Janeiro State University, Rio de Janeiro, Brazil
e-mail: fabricia.fontesdantas@uerj.br

V. Domingues
Neurology Service, Pedro Ernesto University Hospital, Rio de Janeiro State University, Rio de Janeiro, Brazil

G. G. Fernandes
Neuropharmacogenetics Laboratory, Department of Pharmacology and Psychobiology, Roberto Alcântara Gomes Institute Biology (IBRAG), Rio de Janeiro State University, Rio de Janeiro, Brazil

Department of Biophysics and Physiology, Federal University of Juiz de Fora, Minas Gerais, Brazil

M. S. Razzaque (ed.), *Vitamin D Function*, Advances in Experimental Medicine and Biology 1493,
https://doi.org/10.1007/978-3-032-04357-3_4

vitamin D can lead to conditions such as rickets in children and osteoporosis in adults [90]. The fortification of milk with vitamin D in the 1930s effectively eradicated rickets worldwide; however, vitamin D deficiency remains a significant public health concern, highlighting the need for screening in high-risk populations [41, 90]. The factors contributing to vitamin D deficiency include inadequate dietary intake, reduced endogenous synthesis, increased hepatic catabolism, end-organ resistance, and decreased sunlight exposure.

In summary, the synthesis of vitamin D initiates in the skin, where ultraviolet (UV) radiation catalyzes the conversion of 7-dehydrocholesterol into vitamin D3 (cholecalciferol). Once formed, cholecalciferol undergoes hydroxylation via enzymes 25-hydroxylase (CYP2R1 and CYP27A1), leading to its transformation into 25-hydroxyvitamin D3 (calcidiol) in the liver. Subsequently, calcidiol is further hydroxylated to produce 1,25-dihydroxyvitamin D3 (calcitriol) through the action of 1α-hydroxylase (CYP27B1) in renal tissue [15]. This metabolic process is tightly regulated by parathyroid hormone (PTH), which is secreted by the parathyroid glands in response to reductions in serum calcium and phosphate levels (Fig. 1) [54].

Calcitriol not only modulates nongenomic pathways but also functions as a transcriptional regulator of various genes through its interaction with specific receptors, predominantly the nuclear vitamin D receptor (VDR) [20, 38]. VDR acts as a ligand-inducible transcription factor found in various tissues, including the skin, parathyroid glands, and immune cells. Upon binding to calcitriol, it forms a complex with the retinoic acid X receptor (RXR), which regulates the transcription of vitamin D target genes (Fig. 2). The process involves the binding of the VDR to the vitamin D binding protein, heterodimerizing with RXR, and recruiting additional proteins for transcription [38, 56]. The identification of VDR in cell types not directly involved in calcium and phosphorus homeostasis has revealed a plethora of additional biological functions that may be attributed to this molecule.

In addition to its well-established role in the ostheomineral homeostasis, vitamin D is increasingly recognized for its crucial function in modulating the immune system, influencing both innate and adaptive immunity [26, 40]. VDR is expressed on nearly all immune cells, including activated CD4+ and CD8+ T cells, B cells, neutrophils, and antigen-presenting cells such as macrophages and dendritic cells [39, 56, 72]. Vitamin D enhances the antimicrobial capabilities of monocytes and macrophages by promoting the expression of antimicrobial peptides such as cathelicidin and defensins, which disrupt pathogen membranes [61]. It also inhibits pro-inflammatory cytokine production, reducing excessive inflammation—a common characteristic of many autoimmune diseases [1]. Within the adaptive immune system, vitamin D promotes the differentiation of naive T cells into regulatory T cells (Tregs), which are crucial for maintaining immune tolerance and preventing autoimmunity [47]. Additionally, it inhibits the differentiation and proliferation of Th1 and Th17 cells, resulting in reduced production of inflammatory cytokines linked to the pathogenesis of autoimmune diseases [2]. In addition, vitamin D reduces the proliferation and differentiation of B cells into plasma cells, leading to decreased autoantibody production, a mechanism particularly relevant in autoimmune diseases such as systemic lupus erythematosus and rheumatoid arthritis [22]. This expanded understanding of the immune-modulating functions of vitamin D has led to intriguing hypotheses about its potential involvement in the etiology and progression of autoimmune diseases [26, 99].

Calcitriol regulates over 1200 genes in the human genome, and genetic and epigenetic variations further increase its activity. Upregulation of gene transcription may stimulate the release of anti-inflammatory cytokines by immune cells, while downregulation can inhibit these beneficial effects [20, 31]. These immunomodulatory properties have spurred research into how vitamin D deficiency and genetic variations in vitamin D-related pathways may contribute to

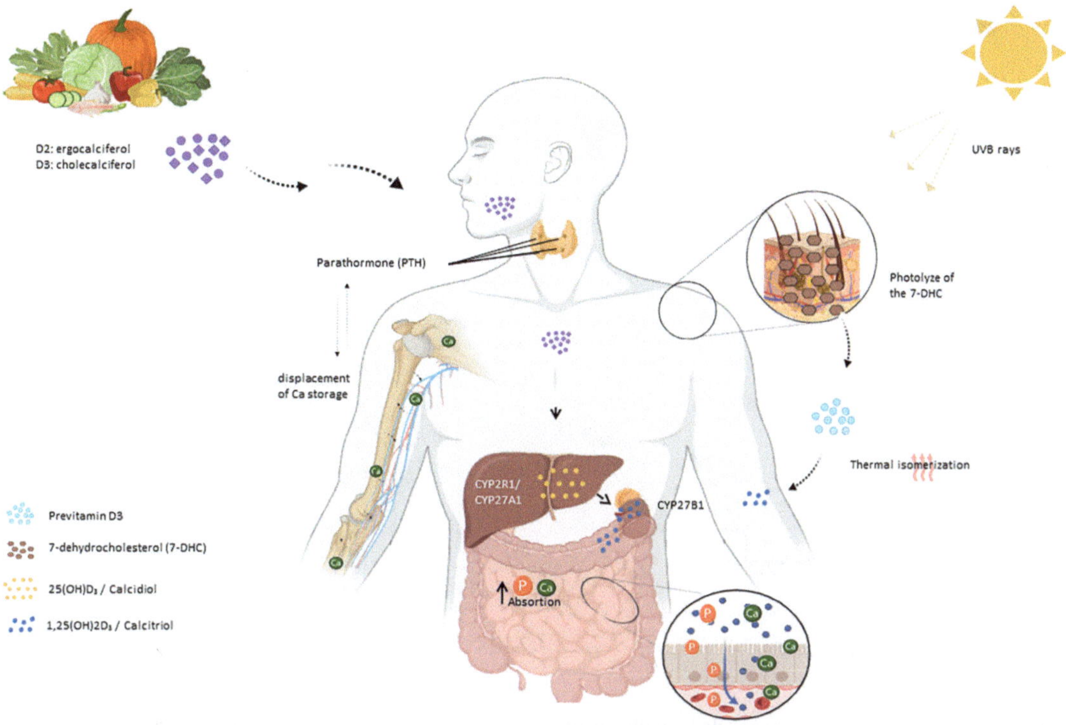

Fig. 1 Photometabolization of vitamin D, competent tissues, and intestinal absorption of calcitriol as the final active molecule. The main source of vitamin D for most humans is exposure to sunlight. During this exposure, UVB radiation photolyzes 7-dehydrocholesterol (7-DHC) in the skin, giving rise to previtamin D3. Afterward, it undergoes thermal isomerization, converting into vitamin D3 itself. Another significant route is dietary intake, where the main sources of vitamin D2 are plants and fungi, while vitamin D3 can be obtained from foods of animal origin. After ingestion, both ergocalciferol and cholecalciferol (D2 and D3) are hydroxylated by 25(OH)D1α hydroxylase enzymes, specifically CYP2R1 and CYP27A1, which are converted into 25(OH)D_3 (calcidiol) in the liver. This protein is transported by vitamin D binding protein (VDBP), and in the kidneys, it is hydroxylated again by CYP27B1, forming 1,25$(OH)_2D_3$ (calcitriol). The final essential mechanism of this process is the regulation of plasma phosphate and calcium levels, which involves a chain of predominantly hormonal molecular events, including those in the liver, kidney, skin, and parathyroid tissues. (This figure was created with BioRender.com)

increased susceptibility to and severity of autoimmune diseases, including multiple sclerosis, type 1 diabetes, and lupus [16, 45, 105]. This chapter focuses on the signaling pathways of vitamin D, its relationship with autoimmune diseases, and the genetic factors involved.

Association of Vitamin D with Major Autoimmune Disorders

Multiple Sclerosis

Multiple sclerosis (MS) is a chronic autoimmune disorder characterized by demyelination of nerve fibers in the central nervous system (CNS) [32]. The incidence of MS is notably greater in countries farther from the equator, which correlates with reduced ultraviolet B (UVB) radiation and consequently lower endogenous vitamin D synthesis [10]. Seasonal variations in MS relapse rates have also been documented, with fewer relapses occurring in the summer months when sunlight exposure and vitamin D levels peak [98]. Research indicates that the month of birth may influence MS risk; individuals born in late autumn, whose mothers are likely to experience greater sun exposure during pregnancy, are at significantly lower risk, whereas those born in late spring, when maternal sun exposure is likely lower, are at increased risk [77].

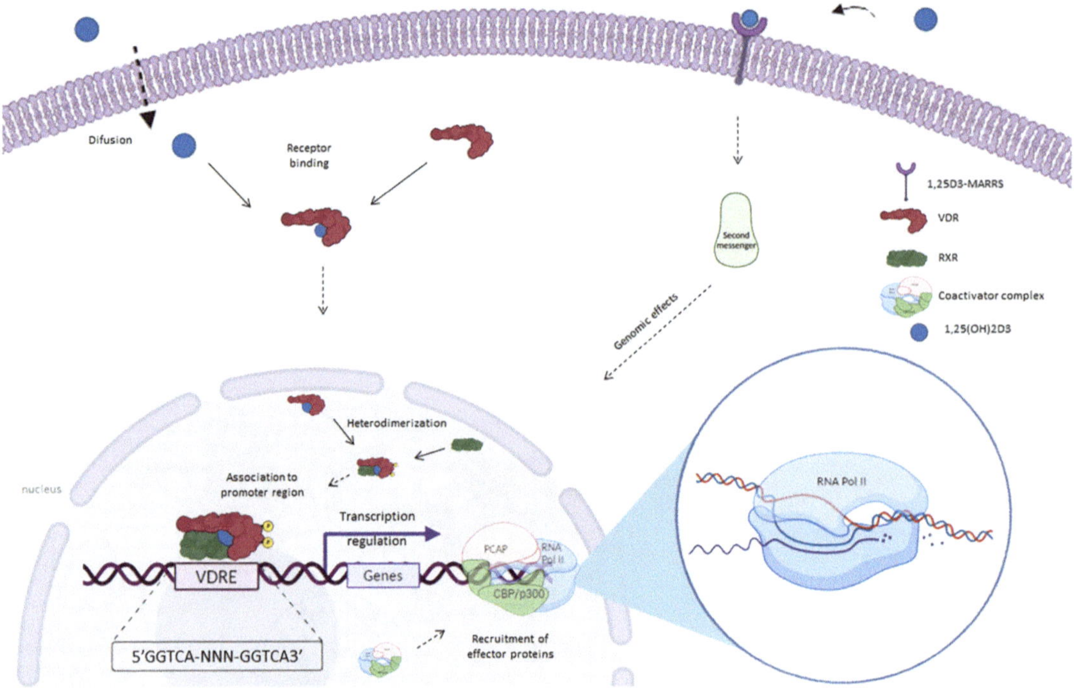

Fig. 2 Cellular pathways associated with active vitamin D, its effectors, and genetic signaling mechanisms. After the conversion of calcidiol to calcitriol, the active form of vitamin D enters the target cell by simple diffusion due to its lipophilic nature. Inside the cell, $1,25(OH)_2D_3$ associates with the vitamin D receptor (VDR), and the resulting complex translocates to the nucleus, where it recruits the retinoic acid receptor (RXR) and acts as an effector complex. This heterodimer shows high affinity for vitamin D responsive elements (VDREs) in the regulatory regions of genes coordinated by $1,25(OH)_2D_3$, regulating the transcription of several genes and recruiting proteins such as RNA polymerase II, which participates in mRNA transcription and subsequent protein synthesis. Additionally, calcitriol can exert nongenomic effects through the membrane-associated rapid-response steroid binding receptor (1,25D3-MARRS). This pathway is characterized by faster onset and involves interactions with secondary mediators, resulting in effects that complement those of the classical genomic pathway. (This figure was created with BioRender.com)

In addition to its known immunomodulatory effects, vitamin D also has neuroprotective properties. It promotes the proliferation and differentiation of oligodendrocytes, enhances neurotrophin expression, reduces the differentiation of microglia toward a pro-inflammatory and neurodegenerative phenotype, mitigates reactive astrogliosis, stabilizes the blood-brain barrier, and decreases oxidative stress [95]. Given the potential role of vitamin D in modulating key pathophysiological aspects of MS, supplementation has been proposed as an adjunctive therapy for MS patients [77]. Although the optimal dosages and durations for vitamin D supplementation are still under investigation, some experts recommend maintaining serum 25(OH)D levels between 40 ng/mL and 60 ng/mL [97, 113].

Systemic Lupus Erythematosus

Systemic lupus erythematosus (SLE) is a multifaceted autoimmune disease characterized by chronic inflammation that can damage organs, including the skin, joints, kidneys, and nervous system [52]. Studies suggest that SLE patients frequently experience vitamin D deficiency, largely due to sun avoidance, chronic use of glucocorticoids, and the presence of kidney disease, all of which reduce vitamin D levels

[66]. Low vitamin D levels are associated not only with more severe disease presentation but also with a higher incidence of cardiovascular risk factors, including arterial thrombosis, elevated LDL cholesterol, triglycerides, BMI, and insulin resistance [65, 83, 115].

A meta-analysis by Irfan et al. [46] highlighted that vitamin D supplementation can reduce the severity of SLE, resulting in improvements in clinical indices such as the Systemic Lupus Erythematosus Disease Activity Index (SLEDAI), the Fatigue Severity Score (FSS), and complement C3 levels [46]. These findings support the view that vitamin D not only impacts musculoskeletal health but also plays a critical role in regulating immune function and inflammation in SLE patients.

The mechanisms through which vitamin D influences SLE involve its interaction with immune cells. Vitamin D is known to modulate both innate and adaptive immune responses by decreasing pro-inflammatory cytokine production and promoting immune tolerance. For example, it helps regulate the activity of dendritic cells and T regulatory cells (Tregs), reducing autoantibody production, which is characteristic of autoimmune diseases such as SLE. Moreover, exogenous vitamin D can counteract some undesirable side effects of glucocorticoids, which are commonly used to manage SLE but can further contribute to vitamin D deficiency and bone loss [66].

Given these connections, maintaining adequate vitamin D levels—typically above 30 ng/mL—is recommended for SLE patients [11]. This approach can help control disease progression and reduce cardiovascular complications, which are major causes of morbidity and mortality in this population. However, further research is needed to determine the optimal dose and frequency of supplementation to maximize its effectiveness in improving the clinical outcomes of patients with SLE.

Rheumatoid Arthritis

Rheumatoid arthritis (RA) is a chronic autoimmune disorder characterized by inflammation of the joints and surrounding tissues, resulting in pain and swelling [101]. Numerous studies have indicated that low levels of vitamin D in RA patients are associated with disease progression [11]. In a cohort study involving 645 patients with early-stage RA, vitamin D deficiency was identified as a potential biomarker for predicting disability and radiographic progression over a 1-year period [68]. RA patients exhibit elevated levels of pro-inflammatory cytokines, such as IL-1β, IFN-γ, and TNF-α. Research has shown that vitamin D, in combination with mediators such as parathyroid hormone, calcium, and phosphorus, plays a role in modulating these inflammatory cytokines in RA [91]. Supplementation with vitamin D in RA patients has been found to provide significant benefits, including pain relief, improved disease activity score 28 (DAS28), and a reduced erythrocyte sedimentation rate (ESR) [36, 69].

Systemic Sclerosis

Systemic sclerosis (SSc), also known as scleroderma, is a complex autoimmune disease characterized by widespread vascular abnormalities, fibrosis of the skin and internal organs, and dysregulation of the immune system [28]. Fibrosis, a hallmark of SSc, involves the excessive deposition of extracellular matrix components, resulting in tissue stiffening and organ dysfunction. Intriguingly, research suggests that the VDR acts as a negative regulator of the TGF-β/Smad signaling pathway. Therefore, impaired VDR signal transduction may contribute to hyperactive TGF-β signaling and abnormal fibroblast activation in SSc [120]. Consistent with this premise, studies have shown that treatment with vitamin D analogs in mice with bleomycin-induced skin fibrosis reduces the extent of fibrosis by downregulating TGF-β signaling [100, 103, 108].

Patients with SSc are more likely to have vitamin D deficiency than the general population is, particularly those with the diffuse cutaneous form of the disease, as opposed to the limited cutaneous form [24, 35]. Furthermore, previous studies have suggested that standard vitamin D supplementation may not adequately correct vitamin D deficiency in SSc patients. This may be due to impaired gastrointestinal absorption in these individuals, indicating that UV light therapy might be more effective than oral supplementation [17, 84, 109]. Groseanu et al. [35] demonstrated that vitamin D supplementation in patients with SSc was associated with a lower frequency of gastrointestinal ulcers than in patients without supplementation, suggesting a potential benefit of vitamin D in managing gastrointestinal complications in SSc patients [35].

Sjogren's Syndrome

Sjogren's syndrome (SS) is a chronic autoimmune disorder characterized by lymphocytic infiltration of the exocrine glands, leading to hallmark symptoms such as dry eyes and dry mouth [49]. A systematic review and meta-analysis by [78] examined 670 SS patients and 857 healthy controls and revealed a high prevalence of hypovitaminosis D in SS patients [78]. In another study, Müller et al. [70] reported reduced levels of 25-hydroxyvitamin D3 (25-OHD3) in SS patients but normal levels of the active metabolite 1,25-dihydroxyvitamin D3, suggesting possible abnormal vitamin D metabolism in these patients, particularly those with high IgM rheumatoid factor levels [70]. Finally, [14] investigated the prevalence of vitamin D deficiency in early-stage primary SS and reported no significant difference in 25-OHD3 levels between SS patients and controls [12]. The connection between vitamin D deficiency and SS remains a topic of debate, as current evidence is insufficient to definitively demonstrate differences in vitamin D levels between SS patients and healthy individuals [60, 123]. However, they reported that leukocytopenia was associated with severe vitamin D deficiency, highlighting a potential subset of SS patients who could benefit from vitamin D supplementation. Despite the inconclusive findings on vitamin D levels, lower vitamin D has been linked to an increased risk of peripheral neuropathy and non-Hodgkin lymphoma in SS patients [4].

Type 1 Diabetes

Type 1 diabetes (T1D) is a chronic autoimmune disorder characterized by the destruction of insulin-producing β-cells in the pancreas, leading to lifelong dependency on exogenous insulin [33]. A comprehensive analysis of the role of vitamin D in T1D revealed significant insights from various studies. Calcitriol improves insulin sensitivity and glycemic control in dexamethasone-induced insulin-resistant rats, highlighting its unique role in regulating glucose homeostasis and insulin secretion through the modulation of calcium influx and intracellular signaling pathways [64]. Another study showed that calcitriol protects pancreatic β-cells from the suppressive effects of interleukin-1β (IL-1β), a cytokine involved in autoimmune insulin-dependent diabetes mellitus. This protection enhances insulin accumulation in isolated rat pancreatic islets and may be linked to the positive impact of vitamin D on mitochondrial calcium metabolism in β-cells [94]. Since β-cell destruction cannot be reversed, vitamin D supplementation is considered more of a preventive than curative measure against this disease [25] (Dankers et al. 2017).

Yu et al. [118] conducted a systematic review demonstrating strong associations between low circulating vitamin D levels and T1D, suggesting that adequate vitamin D status in early life may reduce T1D risk [118]. They highlighted the benefits of vitamin D supplementation in animal models, particularly in NOD mice, and noted potential protective effects in humans, particularly concerning *VDR* polymorphisms and β-cell function. Other meta-analysis studies have revealed an inversely significant association between the circulating 25(OH)D concentration and the risk of T1D and that vitamin D intake during early life is associated with a reduced risk of T1D [29, 43]. In contrast, when a Mendelian randomization approach is used, there is no substantial causal relationship between genetically decreased vitamin D levels and increased T1D risk, indicating that larger studies are needed to explore potential small effects [63].

A recent randomized controlled trial that assessed the combination of saxagliptin and vitamin D in preserving β-cell function in adult-onset T1D reported a significant reduction in insulin dosage and a decrease in β-cell function among participants with increased glutamic acid decarboxylase antibody (GADA) levels [117]. Berti Bastos et al. [14] explored the modulation of T1D through vitamin D and omega-3 polyunsaturated fatty acids and demonstrated potential benefits when these compounds are administered early in the disease course [14]. Collectively, these studies underscore the potential role of vitamin D in T1D, although further investigation is needed to establish definitive causal relationships and therapeutic applications.

Addison's Disease

Addison's disease (AD), also known as primary adrenal insufficiency (PAI), is a rare endocrine disorder characterized by inadequate production of adrenal hormones, primarily cortisol and aldosterone, due to destruction of the adrenal cortex. The condition often arises from autoimmune processes but can also result from infections, cancers, or other diseases that affect the adrenal glands [13]. Vitamin D deficiency and insufficiency are frequently observed in individuals with autoimmune Addison's disease [80, 119]. A retrospective analysis involving 31 patients with AD revealed that 90.3% of these patients had vitamin D concentrations lower than 30 ng/mL, with 19.3% having concentrations below 10 ng/mL [119]. Patients who experienced severe fatigue or limited exercise capacity presented even lower vitamin D levels. A pilot study revealed that high-dose vitamin D therapy significantly increased vitamin D levels and regulated immune cell profiles, including decreased late-activated T cells and increased monocyte numbers, in patients with Addison's disease [75]. However, further studies are needed to determine the optimal dosage, efficacy, and safety of vitamin D supplementation for treating Addison's disease and other endocrine disorders [71].

Crohn's Disease

Crohn's disease (CD) is a relapsing inflammatory bowel disease (IBD) characterized by inflammation throughout the gastrointestinal tract, potentially affecting any area from the mouth to the anus, and can lead to various extraintestinal complications [111]. Recent research suggests that vitamin D deficiency may play a role in the pathogenesis of CD rather than being solely a consequence of intestinal malabsorption [114]. A prospective cohort study involving 72,719 women tracked between 1986 and 2008 identified 122 cases of CD, revealing that higher circulating vitamin D levels significantly reduced the risk of developing the disease [7]. Furthermore, a separate cohort study of 2809 patients with IBD revealed that vitamin D deficiency was associated with an increased risk of colorectal cancer [9]. Adequate vitamin D levels and supplementation in patients with CD and IBD have been linked to reduced reliance on laboratory tests, medications, and radiological services, along with fewer hospitalizations and surgical interventions and improved quality of life scores [8, 50, 79, 93].

Genetic Variability in Vitamin D Signaling

Genetic polymorphisms, or single-nucleotide polymorphisms (SNPs), are hereditary variations in the genetic code present in more than 1% of the human population and can alter the expression or function of the genetic sequence [51]. SNPs located in promoter sequences (genomic regions involved in the initiation of transcription and regulation of gene expression) can alter gene expression levels. A significant proportion of these variants are capable of promoting substantial changes in gene expression [55]. SNPs in promoter-binding regions can either suppress gene expression or influence its rate. The Eukaryotic Promoter Database (EPD) has registered approximately 29,598 human promoter sequences. Most promoter SNPs are located near the transcription start site and can impact biological processes, alter physiological pathways related to specific molecules, and predispose individuals to various pathologies [76]. However, these variants may also play a protective role against other diseases.

The relationships between SNPs in vitamin D signaling genes and autoimmune diseases have been extensively explored. Table 1 summarizes the key SNPs studied, along with relevant data from major databases. According to Ruiz-Ballesteros, multiple polymorphisms in 35 genes involved in the vitamin D signaling pathway may affect the ability to regulate serum vitamin D levels or its systemic effects, depending on the location of these variants [88]. Our research group recently conducted a systematic review on the genetics of vitamin D and autoimmune diseases, integrating findings across various autoimmune diseases. However, we highlight the need for more functional studies to fully understand the impact of these variants on vitamin D signaling [105].

MS has a multifactorial etiology driven by a complex interaction of genetic and epigenetic factors [73]. Numerous studies have identified associations between SNPs in various genes related to vitamin D signaling, including *VDR*, *DHCR7/NADSYN1*, *CYP2R1*, *CYP27A1*, *CYP3A4*, *CYP27B1*, *CYP24A1*, *Megalin-DAB2-Cubilin*, *FGF-23*, and *Klotho*, and the risk of developing MS [96, 105]. However, several authors have reported conflicting results regarding these associations. Susceptibility to developing SLE is also influenced by genetic factors [58]. Some studies have shown that polymorphisms in the *VDR* gene are associated with an increased risk of SLE [67, 92]. On the other hand, one study explored the relationship between genetically determined vitamin D levels and the presence of autoantibodies linked to rheumatoid arthritis (RA) and SLE. A genetic risk score (GRS) for vitamin D was inversely associated with RA autoantibody positivity in individuals at risk, suggesting a protective role of vitamin D-related genes in RA. However, no similar association was found for SLE [110].

Studies have shown that *VDR* polymorphisms—specifically FokI, BsmI, ApaI, and TaqI—are more prevalent in patients with AD than in healthy controls, highlighting the correlation between vitamin D and AD [74]. Furthermore, another study demonstrated that *CYP2A1* expression is significantly reduced in patients with AD, with levels potentially 2.8–3.4 times lower than those in controls [57]. This same study concluded that vitamin D suppresses the inflammatory cytokines CC-ligand chemokine 2 and IL-6.

The *VDBP* SNPs rs4588, rs2060793, and rs2282679 are also correlated with metabolic and autoimmune disorders [37, 105, 112]. Overall, there is an association between low calcidiol levels and the *VDBP* rs4588 CC genotype, which reduces calcitriol status in the context of type 1 diabetes [87]. The rs7041 GG genotype appears to increase the risk of asthma, whereas the rs4588 CA or AA genotypes may play a protective role in asthma progression [27, 87]. However, neurodegenerative diseases are not significantly associated with these variants [59].

Variants in the *VDR* gene can alter protein activity, with more than 14 distinct polymorphisms identified [6, 106]. The most notable *VDR* SNPs include ApaI (rs7975232), BsmI (rs1544410), TaqI (rs731236), and FokI (rs10735810). These silent genetic variants are known to increase mRNA stability [107]. The FokI SNP (rs2228570), also referred to as the start codon polymorphism (SCP), involves the substitution of T (ATG) for C (ACG) in exon 2 [82, 89]. According to Gleba et al., in cancer cells, the shorter variant (424 aa) is more transcriptionally active, whereas the longer variant (427 aa) is less active and less responsive to calcitriol [34]. This study also evaluated the 1,25D3-MARRS receptor and concluded that its expression was directly proportional to the use of the active vitamin D molecule and its effects.

The BsmI SNP (rs1544410), located in intron 8, can alter splice sites for mRNA transcription or affect regulatory elements within the *VDR* intron.

Table 1 Genetic variants involved in vitamin D signaling pathways: number of cataloged SNPs, clinical relevance, and associations with autoimmune diseases

Gene	dbSNP	ClinVar	SNPs related to autoimmune diseases
VDR	25538	Conflicting classifications (27) Benign (40) Likely benign (215) Uncertain significance (182) Likely pathogenic (11) Pathogenic (34)	rs7975232 rs731236 rs2228570 rs1544410 rs10735810 rs11574010 rs11168268 rs2248098 rs1540339 rs4760648 rs3890733 rs11568820 rs12721366 rs2544043 rs4303288 rs11168275 rs757343
CYP2R1	6605	Conflicting classifications (1) Benign (8) Likely benign (63) Uncertain significance (74) Likely pathogenic (5) Pathogenic (30)	rs10741657 rs10766197 rs12369784 rs12794714 rs11023376 rs1257735 rs2060793 rs1993116 rs7116978 rs10500804
WT1	20808	Conflicting classifications (64) Benign (72) Likely benign (547) Uncertain significance (851) Likely pathogenic (51) Pathogenic (109)	rs10767935 rs5030244
MC1R	2613	Conflicting classifications (26) Benign (65) Likely benign (120) Uncertain significance (325) Likely pathogenic (8) Pathogenic (32)	rs1805007 rs1805008 rs1805009 rs2228479 rs885479
CYP27A1	12631	Conflicting classifications (71) Benign (29) Likely benign (432) Uncertain significance (374) Likely pathogenic (110) Pathogenic (171)	rs17470271
CYP27B1	2999	Conflicting classifications (14) Benign (10) Likely benign (231) Uncertain significance (118) Likely pathogenic (30) Pathogenic (68)	rs118204009 rs118204011 rs118204012 rs12368653 rs10877012 rs4646536 rs703842
CYP24A1	12945	Conflicting classifications (16) Benign (73) Likely benign (66) Uncertain significance (135) Likely pathogenic (17) Pathogenic (38)	rs6013897
METTL1	2475	Conflicting classifications (0) Benign (0) Likely benign (5) Uncertain significance (25) Likely pathogenic (3) Pathogenic (7)	rs10877013

(continued)

Table 1 (continued)

Gene	dbSNP	ClinVar	SNPs related to autoimmune diseases
DHCR7	6836	Conflicting classifications (101) Benign (43) Likely benign (359) Uncertain significance (210) Likely pathogenic (164) Pathogenic (148)	rs3829251 rs7944926 rs12785878 rs12800438 rs3794060 rs4945008 rs4944957 rs12785878 rs1790349
Klotho	18993	Conflicting classifications (20) Benign (70) Likely benign (148) Uncertain significance (360) Likely pathogenic (1) Pathogenic (57)	rs1207568 rs9536314
GC	25156	Conflicting classifications (0) Benign (9) Likely benign (6) Uncertain significance (28) Likely pathogenic (9) Pathogenic (20)	rs7041 rs4588 rs12512631 rs2282679 rs3755967 rs17467825 rs2298850
GATA3	15191	Conflicting classifications (7) Benign (60) Likely benign (88) Uncertain significance (138) Likely pathogenic (26) Pathogenic (89)	rs3824662
FOXP3	4525	Conflicting classifications (19) Benign (45) Likely benign (102) Uncertain significance (131) Likely pathogenic (26) Pathogenic (187)	rs3761547 rs3761548
PTPN22	20955	Conflicting classifications (2) Benign (9) Likely benign (16) Uncertain significance (38) Likely pathogenic (1) Pathogenic (11)	rs2476601 rs2488457
PDCD1	4955	Conflicting classifications (0) Benign (18) Likely benign (4) Uncertain significance (19) Likely pathogenic (6) Pathogenic (48)	rs11568821
PTPN2	41115	Conflicting classifications (0) Benign (2) Likely benign (2) Uncertain significance (19) Likely pathogenic (1) Pathogenic (81)	rs1893217 rs478582
RXRB	3794	Conflicting classifications (0) Benign (4) Likely benign (6) Uncertain significance (25) Likely pathogenic (1) Pathogenic (4)	rs1547387

The ApaI SNP (rs7975232), also found in intron 8, does not modify the amino acid sequence but may influence mRNA stability and *VDR* gene expression through linkage disequilibrium (LD) [48]. Its functional impact remains poorly understood [88]. The TaqI SNP (rs731236), located in exon 9 and significantly linked to ApaI, may alter one of the zinc fingers of the nuclear signaling heterodimer that binds to vitamin D response elements (VDREs) in target genes, potentially affecting methylation status [48, 88, 107]. The ApaI variant demonstrated the lowest association with risk factors for autoimmune diseases or vitamin D deficiency [16]. There is a clear indication that ethnicity significantly contributes to the variation in *VDR* variant discrepancies in autoimmune diseases. Additionally, other cosegregating genetic factors, such as the presence of the *HLA-DRB1**15 haplotype, may also influence these outcomes [3, 16, 21].

With respect to RXR, a study by Zhang et al. examined the correlation between the metabolic roles of *RXRα*-rs4842194, rs1045570, and *RXRβ*-rs2076310 SNPs in vitamin D and clinical outcomes in a Chinese cohort infected with HCV [121]. The *RXRβ* rs2076310 genetic variation may influence the transcription and expression of target genes by altering the binding capacity of the VDR-RXR complex, thereby affecting the entire signaling cascade. Another study investigated the rs12004589 and rs9409929 variants in the *RXRα* gene and correlated them with autoimmune outcomes in multiple sclerosis and vitamin D metabolic pathways. No significant associations were detected between alleles or genotypes and clinical or biochemical parameters, nor were there differences compared with the control group [5].

Other polymorphisms in the enzymes responsible for the initial steps of vitamin D conversion are also noteworthy. The first of these enzymes, 25-hydroxylase, is a protein of approximately 500 amino acids (50–55 kDa) synthesized in the liver. It is responsible for metabolizing cholecalciferol and ergocalciferol into calcidiol [86]. Encoded by the *CYP2R1* gene, which is located on the short arm of chromosome 11 at position 15.2 (11p15.2), this enzyme spans 15,500 base pairs. A notable SNP in this gene is rs10741657, characterized by an A > G substitution in the 5′ UTR region [30]. This variant can modulate gene expression by altering translation initiation and stabilizing mRNA transcription, thus impacting the expression of 25-hydroxylase and its enzymatic activity [31, 104]. Notably, a meta-analysis involving 16 studies revealed that the association between *CYP2R1* (rs10741657) and vitamin D deficiency was greater in individuals with the G allele than in those with the A allele in Asian and Caucasian populations [30]. In the study by Khayyatzadeh et al., individuals with the AA genotype for the SNP rs10741657, after supplementation with cholecalciferol, presented serum levels of vitamin D 2.5 times higher than those with the GG genotype [53].

In the kidneys, the enzyme encoded by the *CYP27B1* gene converts calcidiol to calcitriol. Composed of a 507 amino acid (55 kDa) P450 protein, 1α-hydroxylase is located at position 14.1 on the long arm of chromosome 12 (12q14.1). The rs10877012 variant, located in the 3′ UTR, is characterized by a G > T substitution, which may lead to dysregulation of transcript stabilization and cellular targeting [81]. According to a study conducted in a healthy British cohort, there is a connection between the C allele of the *CYP27B1* gene (rs10877012) and reduced serum calcidiol levels [44]. Similarly, Ramos-Lopez et al. reported that in Caucasian women with gestational diabetes, the CC genotype was associated with lower calcidiol levels, and type 1 diabetic individuals with the rs10877012 CC genotype presented reduced CYP27B1 mRNA levels compared with those with the AA genotype [81]. Genetic susceptibility, vitamin D status, and progression of autoimmune disorders suggest a significant influence of the *CYP27B1* rs10877012 polymorphism.

The intricate relationship between genetic polymorphisms, particularly SNPs in the vitamin D signaling pathway, and autoimmune diseases has been extensively explored. While some SNPs appear to predispose individuals to autoimmune diseases, others may have protective effects. Despite the substantial progress in understanding these associations, further functional studies are needed to clarify the precise role of these genetic variants and their potential implications for therapeutic interventions targeting vitamin D pathways.

Gut Microbiota, Autoimmunity, and Vitamin D

Despite the protective effects of vitamin D and its derivatives [19], their efficacy can be influenced by genetic variants, environmental factors such as sun exposure and diet, and the intestinal microbiota. The human microbiome, an ecological community of commensal, symbiotic, and pathogenic microorganisms, inhabits various tissues of the body. Over the past decade, it has become evident that the gut microbiome plays a vital role in shaping the immune system, impacting both health and disease [116]. Disruption of the intestinal barrier can lead to the translocation of gut microorganisms, increasing the serum level of lipopolysaccharide (LPS).

Calcitriol acts on monocytes and macrophages, promoting their transition from pro-inflammatory to tolerogenic states while also modulating the activation status of other immune cells. It facilitates the shift from the M1 phenotype to the M2 phenotype through the nVDR-PPARγ pathway and by upregulating IL-10 expression [102, 122]. To suppress the immune response, calcitriol converts pro-inflammatory T cells into suppressor Th cells by enhancing GATA-3 and c-maf, inhibiting Th1 and Th17 cytokines, and stimulating the production of Th2 cytokines, including IL-9 and IL-13 [23, 62].

In autoimmune diseases, the microbiome is of particular interest because of its ability to trigger immune autoreactivity through molecular mimicry. Metabolites produced by gut microbes, such as short-chain fatty acids (SCFAs), lipids, and vitamins, act as important immunomodulators [18]. SCFAs, for example, are crucial for maintaining the integrity of the blood-brain barrier [18, 116], which may contribute to neurological conditions such as MS.

In immune cells, vitamin D promotes the expression of antimicrobial peptides such as cathelicidins, which help regulate the balance of the commensal population, contributing to reduced microbial diversity. Robles-Vera et al. reported a similar trend in mice with colitis and in healthy mice fed a vitamin D-deficient diet compared with those fed a standard diet [85].

Vitamin D deficiency increases the relative abundance of opportunistic pathogens, which, in the context of intestinal barrier dysfunction, may facilitate bacterial translocation and lead to systemic inflammation. Thus, vitamin D is essential, both directly and indirectly, for maintaining microbiota composition and ensuring the integrity and function of the intestinal barrier.

Conclusions

All stages of vitamin D biochemistry and physiology are significant sites of genetic variation. This includes endogenous biosynthesis, exogenous absorption, further hydroxylation, and activation—processes that involve finely tuned enzymatic reactions closely linked to calcium and phosphorus homeostasis and, as evidence suggests, to the immune response. Although the role of vitamin D in the pathophysiology of various autoimmune diseases is increasingly recognized, research in this area often yields conflicting results. A possible explanation for this inconsistency could be the existence of yet unknown stratified genotypes among autoimmune diseases. Another hypothesis is that, to date, the causal relationship between vitamin D imbalance and the onset of autoimmune diseases remains unclear. Despite these uncertainties, the current understanding of the genetics of vitamin D and autoimmune diseases underscores the urgent need for further research in this field. Additional data could contribute to the development of new therapeutic approaches and potentially support preventive strategies for individuals with a high predisposition to autoimmune conditions.

References

1. Adams JS, Hewison M (2008) Unexpected actions of vitamin D: new perspectives on the regulation of innate and adaptive immunity. Nat Clin Pract Endocrinol Metab 4(2):80–90. Available at: https://doi.org/10.1038/ncpendmet0716
2. Adorini L, Penna G (2008) Control of autoimmune diseases by the vitamin D endocrine system. Nat Clin Pract Rheumatol 4(8):404–412. Available at: https://doi.org/10.1038/ncprheum0855
3. Agliardi C et al (2011) Vitamin D receptor (VDR) gene SNPs influence VDR expression and modulate protection from multiple sclerosis in HLA-DRB1*15-positive individuals. Brain Behav Immun 25(7):1460–1467. Available at: https://doi.org/10.1016/j.bbi.2011.05.015
4. Agmon-Levin N et al (2012) Low levels of vitamin-D are associated with neuropathy and lymphoma among patients with Sjögren's syndrome. J Autoimmun 39(3):234–239. Available at: https://doi.org/10.1016/j.jaut.2012.05.018
5. Agnello L et al (2018) CYP27A1, CYP24A1, and RXR-α polymorphisms, vitamin D, and multiple sclerosis: a pilot study. J Mol Neurosci 66(1):77–84. Available at: https://doi.org/10.1007/s12031-018-1152-9
6. Al-Gharrawi ANR, Anvari E, Fateh A (2023) Association of ApaI rs7975232 and BsmI rs1544410 in clinical outcomes of COVID-19 patients according to different SARS-CoV-2 variants. Sci Rep 13(1):3612. Available at: https://doi.org/10.1038/s41598-023-30859-7
7. Ananthakrishnan AN et al (2012) Higher predicted vitamin D status is associated with reduced risk of Crohn's disease. Gastroenterology 142(3):482–489. Available at: https://doi.org/10.1053/j.gastro.2011.11.040
8. Ananthakrishnan AN et al (2013) Normalization of plasma 25-hydroxy vitamin D is associated with reduced risk of surgery in Crohn's disease. Inflamm Bowel Dis 19(9):1921–1927. Available at: https://doi.org/10.1097/MIB.0b013e3182902ad9
9. Ananthakrishnan AN et al (2014) Association between reduced plasma 25-hydroxy vitamin D and increased risk of cancer in patients with inflammatory bowel diseases. Clin Gastroenterol Hepatol 12(5):821–827. Available at: https://doi.org/10.1016/j.cgh.2013.10.011
10. Ascherio A, Munger KL (2007) Environmental risk factors for multiple sclerosis. Part I: the role of infection. Ann Neurol 61(4):288–299. Available at: https://doi.org/10.1002/ana.21117
11. Athanassiou L et al (2023) Vitamin D and autoimmune rheumatic diseases. Biomolecules 13(4). Available at: https://doi.org/10.3390/biom13040709
12. Baldini C et al (2014) Vitamin D in "early" primary Sjögren's syndrome: does it play a role in influencing disease phenotypes? Rheumatol Int 34(8):1159–1164. Available at: https://doi.org/10.1007/s00296-013-2872-3
13. Barthel A et al (2019) An update on Addison's disease. Exp Clin Endocrinol Diabetes 127(2–03):165–175. Available at: https://doi.org/10.1055/a-0804-2715
14. Bastos TSB, Braga TT, Davanso MR (2022) Vitamin D and omega-3 polyunsaturated fatty acids in type 1 diabetes modulation. Endocr Metab Immune Disord Drug Targets 22(8):815–833. Available at: https://doi.org/10.2174/1871530322666220103114450
15. Bikle DD (2014) Vitamin D metabolism, mechanism of action, and clinical applications. Chem Biol 21(3):319–329. Available at: https://doi.org/10.1016/j.chembiol.2013.12.016
16. Bizzaro G et al (2017) Vitamin D and autoimmune diseases: is vitamin D receptor (VDR) polymorphism the culprit? Isr Med Assoc J 19:438–443
17. Bogh MKB et al (2012) Narrowband ultraviolet B three times per week is more effective in treating vitamin D deficiency than 1600 IU oral vitamin D_3 per day: a randomized clinical trial. Br J Dermatol 167(3):625–630. Available at: https://doi.org/10.1111/j.1365-2133.2012.11069.x
18. Braniste V et al (2009) The gut microbiota influences blood–brain barrier permeability in mice. J Med Sci 17(1):53–55. Available at: https://doi.org/10.1126/scitranslmed.3009759.The
19. Brożyna AA et al (2022) Vitamin D signaling in psoriasis: pathogenesis and therapy. Int J Mol Sci 23(15) Available at: https://doi.org/10.3390/ijms23158575
20. Carlberg C (2023) Genomic signaling of vitamin D. Steroids 198(May):109271. Available at: https://doi.org/10.1016/j.steroids.2023.109271
21. Charoenngam N, Holick MF (2013) Immunologic effects of vitamin D on human health and disease. Am Heart J 26(2):132–143. Available at: file:///C:/Users/User/AppData/Local/Mendeley Ltd./Mendeley Desktop/Downloaded/Charoenngam, Holick - 2020 - Immunologic effects of vitamin d on human health and disease.pdf
22. Chen S et al (2007) Modulatory effects of 1,25-dihydroxyvitamin D3 on human B-cell differentiation. J Immunol 179(3):1634–1647. Available at: https://doi.org/10.4049/jimmunol.179.3.1634
23. Cippitelli M, Santoni A (1998) Vitamin D3: a transcriptional modulator of the interferon-gamma gene. Eur J Immunol 28(10):3017–3030. Available at: https://doi.org/10.1002/(SICI)1521-4141(199810)28:10<3017::AID-IMMU3017>3.0.CO;2-6
24. Corrado A et al (2015) Relationship between body mass composition, bone mineral density, skin fibrosis and 25(OH) vitamin D serum levels in systemic sclerosis. PLoS One 10(9):e0137912. Available at: https://doi.org/10.1371/journal.pone.0137912
25. Dankers W et al (2017) Vitamin D in autoimmunity: molecular mechanisms and therapeutic potential.

Front Immunol 7:697. Available at: https://doi.org/10.3389/fimmu.2016.00697
26. Daryabor G, Gholijani N, Kahmini FR (2023) A review of the critical role of vitamin D axis on the immune system. Exp Mol Pathol 132–133:104866. Available at: https://doi.org/10.1016/j.yexmp.2023.104866
27. Delrue C et al (2024) Investigating vitamin D-binding protein's role in childhood health and development. Int J Mol Sci. Multidisciplinary Digital Publishing Institute (MDPI). Available at: https://doi.org/10.3390/ijms25116272
28. Denton CP, Khanna D (2017) Systemic sclerosis. Lancet 390(10103):1685–1699. Available at: https://doi.org/10.1016/S0140-6736(17)30933-9
29. Dong J-Y et al (2013) Vitamin D intake and risk of type 1 diabetes: a meta-analysis of observational studies. Nutrients 5(9):3551–3562. Available at: https://doi.org/10.3390/nu5093551
30. Duan L et al (2018) Effects of CYP2R1 gene variants on vitamin D levels and status: a systematic review and meta-analysis. Gene:361–369. Elsevier B.V. Available at: https://doi.org/10.1016/j.gene.2018.08.056
31. Fetahu IS, Höbaus J, Kállay E (2014) Vitamin D and the epigenome. Front Physiol 5(April):1–12. Available at: https://doi.org/10.3389/fphys.2014.00164
32. Filippi M et al (2018) Multiple sclerosis. Nat Rev Dis Prim 4(1):43. Available at: https://doi.org/10.1038/s41572-018-0041-4
33. Gillespie KM (2006) Type 1 diabetes: pathogenesis and prevention. Can Med Assoc J 175(2):165–170. Available at: https://doi.org/10.1503/cmaj.060244
34. Gleba JJ et al (2022) Polymorphism of VDR gene and the sensitivity of human leukemia and lymphoma cells to active forms of vitamin D. Cancers 14(2). Available at: https://doi.org/10.3390/cancers14020387
35. Groseanu L et al (2016) Low vitamin D status in systemic sclerosis and the impact on disease phenotype. Eur J Rheumatol 3(2):50–55. Available at: https://doi.org/10.5152/eurjrheum.2015.0065
36. Guan Y et al (2020) The effect of vitamin d supplementation on rheumatoid arthritis patients: a systematic review and meta-analysis. Front Med 7:596007. Available at: https://doi.org/10.3389/fmed.2020.596007
37. Haldar D et al (2018) Association of VDBP and CYP2R1 gene polymorphisms with vitamin D status in women with polycystic ovarian syndrome: a north Indian study. Eur J Nutr 57(2):703–711. Available at: https://doi.org/10.1007/s00394-016-1357-z
38. Haussler MR et al (1998) The nuclear vitamin D receptor: biological and molecular regulatory properties revealed. J Bone Miner Res 13(3):325–349. Available at: https://doi.org/10.1359/jbmr.1998.13.3.325
39. Heine G et al (2008) 1,25-dihydroxyvitamin D (3) promotes IL-10 production in human B cells. Eur J Immunol 38(8):2210–2218. Available at: https://doi.org/10.1002/eji.200838216
40. Hewison M (2012) Vitamin D and immune function: an overview. Proc Nutr Soc 71(1):50–61. Available at: https://doi.org/10.1017/S0029665111001650
41. Holick MF (2006) Resurrection of vitamin D deficiency and rickets. J Clin Invest:2062–2072. Available at: https://doi.org/10.1172/JCI29449
42. Holtrop ME et al (1986) Effects of serum calcium and phosphorus on skeletal mineralization in vitamin D-deficient rats. Am J Phys 251(2 Pt 1):E234–E240. Available at: https://doi.org/10.1152/ajpendo.1986.251.2.E234
43. Hou Y et al (2021) A dose–response meta-analysis between serum concentration of 25-hydroxy vitamin D and risk of type 1 diabetes mellitus. Eur J Clin Nutr 75(7):1010–1023. Available at: https://doi.org/10.1038/s41430-020-00813-1
44. Hyppönen E et al (2009) Serum 25-hydroxyvitamin D and IgE–a significant but nonlinear relationship. Allergy 64(4):613–620. Available at: https://doi.org/10.1111/j.1398-9995.2008.01865.x
45. Illescas-Montes R et al (2019) Vitamin D and autoimmune diseases. Life Sci 233:116744. Available at: https://doi.org/10.1016/j.lfs.2019.116744
46. Irfan SA et al (2022) Effects of vitamin D on systemic lupus erythematosus disease activity and autoimmunity: a systematic review and meta-analysis. Cureus 14(6):e25896. Available at: https://doi.org/10.7759/cureus.25896
47. Jeffery LE et al (2009) 1,25-Dihydroxyvitamin D3 and IL-2 combine to inhibit T-cell production of inflammatory cytokines and promote development of regulatory T cells expressing CTLA-4 and FoxP3. J Immunol 183(9):5458–5467. Available at: https://doi.org/10.4049/jimmunol.0803217
48. Jeong SP, Sharma N, An SSA (2024) Role of calcitriol and vitamin D receptor (VDR) gene polymorphisms in Alzheimer's disease. Int J Mol Sci. Multidisciplinary Digital Publishing Institute (MDPI). Available at: https://doi.org/10.3390/ijms25094806
49. Jonsson R et al (2018) Current concepts on Sjögren's syndrome–classification criteria and biomarkers. Eur J Oral Sci 126(Suppl 1):37–48. Available at: https://doi.org/10.1111/eos.12536
50. Jørgensen SP et al (2010) Clinical trial: vitamin D3 treatment in Crohn's disease–a randomized double-blind placebo-controlled study. Aliment Pharmacol Ther 32(3):377–383. Available at: https://doi.org/10.1111/j.1365-2036.2010.04355.x
51. Karki R et al (2015) Defining "mutation" and "polymorphism" in the era of personal genomics. BMC Med Genet 8:37. Available at: https://doi.org/10.1186/s12920-015-0115-z

52. Kaul A et al (2016) Systemic lupus erythematosus. Nat Rev Dis Prim 2(1):16039. Available at: https://doi.org/10.1038/nrdp.2016.39
53. Khayyatzadeh SS et al (2019) A variant in CYP2R1 predicts circulating vitamin D levels after supplementation with high-dose of vitamin D in healthy adolescent girls. J Cell Physiol 234(8):13977–13983. Available at: https://doi.org/10.1002/jcp.28083
54. Khundmiri SJ, Murray RD, Lederer E (2016) PTH and Vitamin D. Compr Physiol 6(2):561–601. Available at: https://doi.org/10.1002/cphy.c140071
55. Knight JC (2005) Regulatory polymorphisms underlying complex disease traits. J Mol Med (Berl) 83(2): 97–109. Available at: https://doi.org/10.1007/s00109-004-0603-7
56. Kongsbak M et al (2013) The vitamin d receptor and T-cell function. Front Immunol 4:148. Available at: https://doi.org/10.3389/fimmu.2013.00148
57. Kraus AU et al (2018) Vitamin D effects on monocytes' CCL-2, IL6 and CD14 transcription in Addison's disease and HLA susceptibility. J Steroid Biochem Mol Biol 177:53–58. Available at: https://doi.org/10.1016/j.jsbmb.2017.07.026
58. Kwon Y-C et al (2019) Update on the genetics of systemic lupus erythematosus: genome-wide association studies and beyond. Cells 8(10). Available at: https://doi.org/10.3390/cells8101180
59. Langer-Gould A et al (2018) Vitamin d-binding protein polymorphisms, 25-hydroxyvitamin d, sunshine and multiple sclerosis. Nutrients 10(2). Available at: https://doi.org/10.3390/nu10020184
60. Li L, Chen J, Jiang Y (2019) The association between vitamin D level and Sjögren's syndrome: A meta-analysis. Int J Rheum Dis:532–533. Available at: https://doi.org/10.1111/1756-185X.13474
61. Liu PT et al (2006) Toll-like receptor triggering of a vitamin D-mediated human antimicrobial response (Science) 311(5768):1770–1773. Available at: https://doi.org/10.1126/science.1123933
62. Mahon BD et al (2003) The targets of vitamin D depend on the differentiation and activation status of CD4 positive T cells. J Cell Biochem 89(5): 922–932. Available at: https://doi.org/10.1002/jcb.10580
63. Manousaki D et al (2021) Vitamin D levels and risk of type 1 diabetes: a Mendelian randomization study. PLoS Med 18(2):e1003536. Available at: https://doi.org/10.1371/journal.pmed.1003536
64. Mendes AKB et al (2022) 1α,25-(OH)2 vitamin D3 prevents insulin resistance and regulates coordinated exocytosis and insulin secretion. J Nutr Biochem 99: 108864. Available at: https://doi.org/10.1016/j.jnutbio.2021.108864
65. Mok CC (2006) Accelerated atherosclerosis, arterial thromboembolism, and preventive strategies in systemic lupus erythematosus. Scand J Rheumatol 35(2):85–95. Available at: https://doi.org/10.1080/03009740600601526
66. Mok CC (2013) Vitamin D and systemic lupus erythematosus: an update. Expert Rev Clin Immunol 9(5):453–463. Available at: https://doi.org/10.1586/eci.13.19
67. Monticielo OA et al (2012) Vitamin D and polymorphisms of VDR gene in patients with systemic lupus erythematosus. Clin Rheumatol 31(10): 1411–1421. Available at: https://doi.org/10.1007/s10067-012-2021-5
68. Mouterde G et al (2020) Association between vitamin D deficiency and disease activity, disability, and radiographic progression in early rheumatoid arthritis: the ESPOIR cohort. J Rheumatol 47(11): 1624–1628. Available at: https://doi.org/10.3899/jrheum.190795
69. Mukherjee D et al (2019) Effect of 1,25 dihydroxy vitamin D3 supplementation on pain relief in early rheumatoid arthritis. J Family Med Prim Care 8(2): 517–522. Available at: https://doi.org/10.4103/jfmpc.jfmpc_446_18
70. Müller K et al (1990) Abnormal vitamin D3 metabolism in patients with primary Sjögren's syndrome. Ann Rheum Dis 49(9):682–684. Available at: https://doi.org/10.1136/ard.49.9.682
71. Muscogiuri G et al (2014) Mechanisms in endocrinology: vitamin D as a potential contributor in endocrine health and disease. Eur J Endocrinol 171(3): R101–R110. Available at: https://doi.org/10.1530/EJE-14-0158
72. Nagy L et al (2012) Nuclear hormone receptors enable macrophages and dendritic cells to sense their lipid environment and shape their immune response. Physiol Rev 92(2):739–789. Available at: https://doi.org/10.1152/physrev.00004.2011
73. Oksenberg JR et al (2008) The genetics of multiple sclerosis: SNPs to pathways to pathogenesis. Nat Rev Genet 9(7):516–526. Available at: https://doi.org/10.1038/nrg2395
74. Pani MA et al (2002) Vitamin D receptor genotype is associated with Addison's disease. Eur J Endocrinol 147(5):635–640. Available at: https://doi.org/10.1530/eje.0.1470635
75. Penna-Martinez M et al (2018) High-dose vitamin D in Addison's disease regulates T cells and monocytes: a pilot trial. Nutrition 49:66–73. Available at: https://doi.org/10.1016/j.nut.2017.10.021
76. Périer RC et al (2000) The eukaryotic promoter database (EPD). Nucleic Acids Res 28(1):302–303. Available at: https://doi.org/10.1093/nar/28.1.302
77. Pierrot-Deseilligny C, Souberbielle J-C (2017) Vitamin D and multiple sclerosis: an update. Mult Scler Relat Disord 14:35–45. Available at: https://doi.org/10.1016/j.msard.2017.03.014
78. Radić M et al (2023) Vitamin D and Sjögren's disease: revealing the connections-a systematic review and meta-analysis. Nutrients 15(3) Available at: https://doi.org/10.3390/nu15030497

79. Raftery T et al (2015) Effects of vitamin D supplementation on intestinal permeability, cathelicidin and disease markers in Crohn's disease: results from a randomized double-blind placebo-controlled study. United European Gastroenterol J 3(3):294–302. Available at: https://doi.org/10.1177/2050640615572176
80. Ramagopalan SV et al (2013) Hospital admissions for vitamin D related conditions and subsequent immune-mediated disease: record-linkage studies. BMC Med 11:171. Available at: https://doi.org/10.1186/1741-7015-11-171
81. Ramos-Lopez E et al (2008) Gestational diabetes mellitus and vitamin D deficiency: genetic contribution of CYP27B1 and CYP2R1 polymorphisms. Diabetes Obes Metab:683–685. Blackwell Publishing Ltd. Available at: https://doi.org/10.1111/j.1463-1326.2008.00879.x
82. Raza S et al (2019) Molecular mechanism of cancer susceptibility associated with Fok1 single-nucleotide polymorphism of VDR in relation to breast cancer. Asian Pac J Cancer Prev 20(1):199–206. Available at: https://doi.org/10.31557/APJCP.2019.20.1.199
83. Reynolds JA et al (2012) 25-Hydroxyvitamin D deficiency is associated with increased aortic stiffness in patients with systemic lupus erythematosus. Rheumatology 51(3):544–551. Available at: https://doi.org/10.1093/rheumatology/ker352
84. Robberecht E et al (2011) Sunlight is an important determinant of vitamin D serum concentrations in cystic fibrosis. Eur J Clin Nutr 65(5):574–579. Available at: https://doi.org/10.1038/ejcn.2010.280
85. Robles-Vera I et al (2019) Impact of vitamin D deficit on the rat gut microbiome. Nutrients 11-(11) Available at: https://doi.org/10.3390/nu11112564
86. Roff A, Wilson RT (2008) A novel SNP in a vitamin D response element of the CYP24A1 promoter reduces protein binding, transactivation, and gene expression. J Steroid Biochem Mol Biol 112(1–3):47–54. Available at: https://doi.org/10.1016/j.jsbmb.2008.08.009
87. Rozmus D et al (2022) Rs7041 and rs4588 polymorphisms in vitamin D binding protein-encoding gene (VDBP) and the risk of diseases. Int J Mol Sci. MDPI. Available at: https://doi.org/10.3390/ijms23020933
88. Ruiz-Ballesteros AI et al (2020) Association of vitamin D metabolism gene polymorphisms with autoimmunity: evidence in population genetic studies. Int J Mol Sci 21(24):1–24. Available at: https://doi.org/10.3390/ijms21249626
89. Saccone D, Asani F, Bornman L (2015) Regulation of the vitamin D receptor gene by environment, genetics and epigenetics. Gene:171–180. Elsevier B.V. Available at: https://doi.org/10.1016/j.gene.2015.02.024
90. Sahay M, Sahay R (2012) Rickets-vitamin D deficiency and dependency. Indian J Endocrinol Metab 16(2):164–176. Available at: https://doi.org/10.4103/2230-8210.93732
91. Sakyi SA et al (2022) Profiling vitamin D, its mediators and proinflammatory cytokines in rheumatoid arthritis: a case–control study. Immun Inflamm Dis 10(8):e676. Available at: https://doi.org/10.1002/iid3.676
92. Salimi S et al (2019) Vitamin D receptor rs2228570 and rs731236 polymorphisms are susceptible factors for systemic lupus erythematosus. Adv Biomed Res 8:48. Available at: https://doi.org/10.4103/abr.abr_19_19
93. Samson CM et al (2012) Improved outcomes with quality improvement interventions in pediatric inflammatory bowel disease. J Pediatr Gastroenterol Nutr 55(6):679–688. Available at: https://doi.org/10.1097/MPG.0b013e318262de16
94. Sandler S, Buschard K, Bendtzen K (1994) Effects of 1,25-dihydroxyvitamin D3 and the analogs MC903 and KH1060 on interleukin-1β-induced inhibition of rat pancreatic islet β-cell function in vitro. Immunol Lett 41(1):73–77. Available at: https://doi.org/10.1016/0165-2478(94)90059-0
95. Sangha A et al (2023) The role of vitamin D in neuroprotection in multiple sclerosis: an update. Nutrients 15(13) Available at: https://doi.org/10.3390/nu15132978
96. Scazzone C et al (2021) Vitamin D and genetic susceptibility to multiple sclerosis. Biochem Genet 59(1):1–30. Available at: https://doi.org/10.1007/s10528-020-10010-1
97. Shoemaker TJ, Mowry EM (2018) A review of vitamin D supplementation as disease-modifying therapy. Mult Scler 24(1):6–11. Available at: https://doi.org/10.1177/1352458517738131
98. Simpson SJ et al (2011) Latitude is significantly associated with the prevalence of multiple sclerosis: a meta-analysis. J Neurol Neurosurg Psychiatry 82(10):1132–1141. Available at: https://doi.org/10.1136/jnnp.2011.240432
99. Sîrbe C et al (2022) An update on the effects of vitamin D on the immune system and autoimmune diseases. Int J Mol Sci 23(17) Available at: https://doi.org/10.3390/ijms23179784
100. Slominski A et al (2013) 20S-hydroxyvitamin D3, noncalcemic product of CYP11A1 action on vitamin D3, exhibits potent antifibrogenic activity in vivo. J Clin Endocrinol Metab 98(2):E298–E303. Available at: https://doi.org/10.1210/jc.2012-3074
101. Smolen JS, Aletaha D, McInnes IB (2016) Rheumatoid arthritis. Lancet 388(10055):2023–2038. Available at: https://doi.org/10.1016/S0140-6736(16)30173-8
102. Stachowicz-Suhs M et al (2024) Calcitriol promotes M2 polarization of tumor-associated macrophages in 4T1 mouse mammary gland cancer via the induction of proinflammatory cytokines. Sci Rep 14(1):3778.

Available at: https://doi.org/10.1038/s41598-024-54433-x
103. Terao M et al (2015) A vitamin D analog inhibits Th2 cytokine- and TGFβ -induced periostin production in fibroblasts: a potential role for vitamin D in skin sclerosis. Dermatoendocrinol 7(1):e1010983. Available at: https://doi.org/10.1080/19381980.2015.1010983
104. Thacher TD et al (2015) CYP2R1 mutations impair generation of 25-hydroxyvitamin D and cause an atypical form of vitamin D deficiency. J Clin Endocrinol Metabol 100(7):E1005–E1013. Available at: https://doi.org/10.1210/jc.2015-1746
105. Trefilio LM et al (2024) The impact of genetic variants related to vitamin D and autoimmunity: a systematic review. Heliyon 10(7):e27700. Available at: https://doi.org/10.1016/j.heliyon.2024.e27700
106. Uitterlinden AG et al (2004) Genetics and biology of vitamin D receptor polymorphisms. Gene 338(2): 143–156. Available at: https://doi.org/10.1016/j.gene.2004.05.014
107. Usategui-Martín R et al (2022) Vitamin D receptor (VDR) gene polymorphisms modify the response to vitamin D supplementation: a systematic review and meta-analysis. Nutrients 14(2) Available at: https://doi.org/10.3390/nu14020360
108. Usategui A et al (2014) Topical vitamin D analog calcipotriol reduces skin fibrosis in experimental scleroderma. Arch Dermatol Res 306(8):757–761. Available at: https://doi.org/10.1007/s00403-014-1466-6
109. Vacca A et al (2009) Vitamin D deficiency and insufficiency in 2 independent cohorts of patients with systemic sclerosis. J Rheumatol 36(9):1924–1929. Available at: https://doi.org/10.3899/jrheum.081287
110. Vanderlinden LA et al (2022) Relationship between a vitamin D genetic risk score and autoantibodies among first-degree relatives of probands with rheumatoid arthritis and systemic lupus erythematosus. Front Immunol 13:881332. Available at: https://doi.org/10.3389/fimmu.2022.881332
111. Veauthier B, Hornecker JR (2018) Crohn's disease: diagnosis and management. Am Fam Physician 98(11):661–669
112. Voltan G et al (2023) Vitamin D: An overview of gene regulation, ranging from metabolism to genomic effects. Genes 14(9). Available at: https://doi.org/10.3390/genes14091691
113. Ward M, Goldman MD (2022) Epidemiology and pathophysiology of multiple sclerosis. Continuum (Minneap Minn) 28(4):988–1005. Available at: https://doi.org/10.1212/CON.0000000000001136
114. White JH (2018) Vitamin D deficiency and the pathogenesis of Crohn's disease. J Steroid Biochem Mol Biol 175:23–28. Available at: https://doi.org/10.1016/j.jsbmb.2016.12.015
115. Wu PW et al (2009) 25-hydroxyvitamin D and cardiovascular risk factors in women with systemic lupus erythematosus. Arthritis Rheum 61(10): 1387–1395. Available at: https://doi.org/10.1002/art.24785
116. Yamamoto EA, Jørgensen TN (2020) Relationships between vitamin D, gut microbiome, and systemic autoimmunity. Front Immunol 10(January):1–13. Available at: https://doi.org/10.3389/fimmu.2019.03141
117. Yan X et al (2023) Combination therapy with saxagliptin and vitamin D for the preservation of β-cell function in adult-onset type 1 diabetes: a multicenter, randomized, controlled trial. Signal Transduct Target Ther 8(1):158. Available at: https://doi.org/10.1038/s41392-023-01369-9
118. Yu J et al (2022) Vitamin D and Beta cells in type 1 diabetes: a systematic review. Int J Mol Sci 23(22). Available at: https://doi.org/10.3390/ijms232214434
119. Zawadzka K et al (2021) Vitamin D status and its associations with clinical and laboratory parameters in patients with Addison's disease. Folia Med Cracov 61(2):65–78. Available at: https://doi.org/10.24425/fmc.2021.137224
120. Zerr P et al (2015) Vitamin D receptor regulates TGF-β signaling in systemic sclerosis. Ann Rheum Dis 74(3):e20. Available at: https://doi.org/10.1136/annrheumdis-2013-204378
121. Zhang R et al (2021) Genetic variant of RXR involved in the vitamin D metabolic pathway was linked to HCV infection outcomes among a high-risk Chinese population. Infect Genet Evol 87. Available at: https://doi.org/10.1016/j.meegid.2020.104641
122. Zhang Y et al (2012) Vitamin D inhibits monocyte/macrophage proinflammatory cytokine production by targeting MAPK phosphatase-1. J Immunol 188(5): 2127–2135. Available at: https://doi.org/10.4049/jimmunol.1102412
123. Zhao M et al (2023) Serum vitamin D levels and Sjogren's syndrome: bidirectional Mendelian randomization analysis. Arthritis Res Ther 25(1):79. Available at: https://doi.org/10.1186/s13075-023-03062-2

Vitamin D and Brain Health

Rahnuma Ahmad, Kona Chowdhury, Susmita Sinha, and Mainul Haque

Abstract

Vitamin D is widely known for its influence on the homeostasis of calcium and phosphate levels in the body and its involvement in maintaining musculoskeletal health. In its active form, 1,25-dihydroxy cholecalciferol, this vitamin has immunomodulatory effects on both the innate and adaptive immune systems. The active form of vitamin D plays a neuroprotective role in the central nervous system, highlighting its importance in maintaining brain health. Studies have noted the role of vitamin D in the formation of antioxidants, reduction of pro-inflammatory cytokines, lowering of oxidative stress, improvement of mitochondrial respiratory health, and slowing of the progression of neuronal damage. This vitamin may also influence development by facilitating the differentiation of cells, the expression of neurotrophic factors, the regulation of cytokines, the synthesis of neurotransmitters, intracellular calcium signaling, and the expression of genes involved in neuronal differentiation. An association has been noted between vitamin D deficiency and the risk of developing several neuronal diseases, including Parkinson's disease, headaches, multiple sclerosis, and Alzheimer's disease. The neuroprotective role of vitamin D suggests the potential benefit of vitamin supplementation in slowing the progression of such conditions and promoting brain health. It is essential to conduct further studies in the future to better understand the mechanisms by which vitamins affect brain health. It is also necessary to determine the potential benefits of vitamin D supplementation in patients with neurological diseases.

Keywords

Function · Participation · Calciferol · Cholecalciferol · Ergocalciferol · Viosterol · Keep in existence · Conserve · Succeeded in keeping · Narrative review

R. Ahmad
Department of Physiology, Medical College for Women and Hospital, Dhaka, Bangladesh

Scientific Committee, Global Alliance for Infections in Surgery, Macerata, Italy

K. Chowdhury
Department of Pediatrics, Enam Medical College and Hospital, Dhaka, Bangladesh

S. Sinha
Department of Physiology, Enam Medical College and Hospital, Dhaka, Bangladesh

M. Haque (✉)
Department of Pharmacology and Therapeutics, Eastern Medical College and Hospital, Cumilla, Bangladesh

Scientific Committee, Global Alliance for Infections in Surgery, Macerata, Italy

Department of Research Karnavati School of Dentistry, Karnavati University, Gandhi Nagar, Gujarat, India
e-mail: runurono@gmail.com

M. S. Razzaque (ed.), *Vitamin D Function*, Advances in Experimental Medicine and Biology 1493,
https://doi.org/10.1007/978-3-032-04357-3_5

Introduction

In 1922, McCollum and colleagues discovered the presence of vitamin D, a crucial component in the treatment of painful pediatric bone disease known as rickets [1–3]. Patients with rickets are highly deficient in nutrients, including vitamin D [4]. Certain foods, e.g., fish-liver oils and fatty fish, are the paramount sources of this micronutrient. Additionally, meat, beef liver, offal, egg yolks, dairy products, cheese, and mushrooms contain vitamin D in the form of vitamin D3 and its metabolite 25-hydroxycholecalciferol [5, 6]. Vitamin D is also recognized as a "sunshine vitamin" [7]. Approximately 90% of the vitamin D supply occurs through dermal synthesis. The human skin contains 7-dehydrocholesterol, and when we are exposed to sunshine, this cholesterol absorbs ultraviolet B (UVB) radiation, and 7-dehydrocholesterol is transformed into previtamin D3, which successively isomerizes into 25-hydroxycholecalciferol in the liver and then converts it to its most active form, 1,25-dihydroxycholecalciferol, in the kidney with the aid of the 1-α-hydroxylase enzyme [7–9]. Therefore, it has been recommended that people be exposed to sunlight for 8 min to 2 h daily in spring and winter, especially in cold countries [10]. The activated form of vitamin D is also a steroid hormone that is essential for intestinal calcium ion absorption, homeostasis, and bone health [11, 12].

Vitamin D3 activates all nucleated cells' vitamin D receptors [13]. In addition to the small intestine, vitamin D receptors are also present in the colon, mononuclear cells, osteoblasts, beta cells of the islets of Langerhans, activated T and B lymphocytes, and essential organs such as the brain, heart, kidneys, skin, breasts, and prostate gland [14]. Its primary physiological purpose is facilitating calcium homeostasis by promoting intestinal calcium absorption and enterocyte differentiation [15].

Vitamin D Beyond Extraskeletal Functions

The role of vitamin D in the human health extends beyond that commonly perceived. This vitamin is needed for several extraskeletal systems, and this segment of the chapter attempts to elaborate on certain activities of vitamin D. Vitamin D and its active metabolite 1,25-dihydroxycholecalciferol protect against and promote corneal epithelial fencing [16]. Additionally, vitamin D possesses anti-inflammatory pharmacodynamic properties [17] and helps minimize eye surface-related inflammatory diseases, e.g., dry eye disease and keratoconus [18]. The anti-inflammatory effects are mediated by inhibiting the synthesis of pro-inflammatory molecules, such as tumor necrosis factor-α, and minimizing the migration of Langerhans cells [19]. The anti-inflammatory properties of vitamin D also reduce the eye-aging process [20]. Nevertheless, deficiency promotes "posterior subcapsular cataracts and early age-related cataracts" [21]. Vitamin D remodels the corneal extracellular matrix and promotes the lesion restorative process [18].

Vitamin D has a beneficial effect on cardiac health [22]. Insufficiency of this vitamin often leads to an increase in the incidence of cardiovascular diseases such as essential hypertension, ischemic heart disease, congestive heart failure, and cardiac failure and may result in a high mortality rate [22–25], whereas overdose increases the possibility of adverse drug reactions, e.g., vomiting, encephalopathy, polyuria, polydipsia, and azotaemia [26]. The vitamin D receptor is present in all core cardiac, vascular, and endothelial cell types, including arterial wall cells, platelets, cardiomyocytes, dendritic cells, macrophages, and other immune cells [27–29]. Biologically active vitamin D binds to its receptor, which controls various genes involved in primary processes possibly applicable to cardiovascular illness, such as membrane transport, matrix homeostasis, cell adhesion, cell proliferation and differentiation, apoptosis, oxidative stress, and adaptive and innate immunity [30–32]. Furthermore, vitamin D maintains normal blood pressure through the actions of smooth muscle and endothelial cells [22, 33]. Endothelial cells function as a fencing mechanism between blood and tissues. These cells are involved in the regulation and coordination of inflammation [34, 35], angiogenesis [36, 37], vasomotion [38], and hemostasis [39, 40]. Arterial thrombosis can be induced through vascular smooth muscle cell-derived tissue factor, which is inhibited by vitamin D [22, 41]. Nonetheless, the addition of synthetically derived active vitamin D (calcitriol) and an analog of vitamin D (paricalcitol) flattened the tumor necrosis factor-alpha-stimulated tissue factor countenance [42, 43] (2.01 ± 0.24 and 1.32 ± 0.14 vs. 3.02 ± 0.39 pmol/mg protein, $p < 0.05$) [22]. Hence, it promotes cell and tissue responses toward the nuclear factor-kappa B communication system, and proteinase-activated receptor 2 activation reinstates typical vitamin D receptor availability and improves the tissue factor blocking effect [22, 42, 44].

Vitamin D preserves renal health [45]. Upon deficiency, renal physiological function is impaired [46]. Vitamin D inhibits the transforming growth factor-alpha and epidermal growth factor receptor progression route [47] and reduces the inflammatory process [48, 49], which leads to apoptosis [50] and fibrosis [51] in renal tissue. Wu et al. [52] revealed that the urine protein/creatinine ratio was considerably lower among vitamin D (experimental) and control clusters at 56, 112, and 168 days ($p < 0.05$). The experimental group received a "vitamin D + angiotensin receptor blocker or angiotensin-converting enzyme inhibitor" and a control (angiotensin receptor blocker or angiotensin-converting enzyme inhibitor) cluster. An optimistic and sanguine relationship has been observed between a reduction in the urine protein/creatinine ratio at 168 days and the starting point of the 25-hydroxyvitamin D serum level in the vitamin D cluster ($r = 0.738$, $p < 0.001$) [52]. Minimum-dose active vitamin D supplementation has a potential therapeutic effect on alleviating albuminuria in patients with chronic kidney disease, even those with a serum 25-hydroxyvitamin D concentration above 50 nmol/L [28, 52–54]. Furthermore, patients with progressive chronic kidney disease begin synthesizing high levels of fibroblast growth factor 23 [55]. Fibroblast growth factor 23 directly inhibits 25-hydroxyvitamin D3 1-alpha-hydroxylase, which is responsible for converting vitamin D to its active form; therefore, the synthesis of an adequate amount of 1,25-vitamin D (active form) fails [56], which ultimately causes further damage to renal function [57]. An overdose of vitamin D increases the possibility of hypercalcemia [58] and extraosseous calcification [59, 60] and further deteriorates the existing condition of chronic kidney disease patients [61, 62].

The liver is the principal location of vitamin D production [63]. Thus, most patients with poor hepatic function are deficient in vitamin D [63, 64]. Vitamin D insufficiency is related to the growth and progression of hepatic disorders, e.g., chronic hepatitis B [65, 66], chronic hepatitis C [67, 68], and nonalcoholic fatty liver disease [69–71]. Vitamin D possesses anti-inflammatory, antifibrotic, antisteatotic, and metabolic properties that provide a conceivable mode of action by which vitamin D influences the avert of the diverse steps of the nonalcoholic fatty liver disease sequence and fatal clinical outcomes [72, 73].

We have already discussed the pleotropic functions of vitamin D on extraskeletal vital organs. The primary target of the write-up is the role of vitamin D in the brain and neurological function, as well as potential disease prevention.

Vitamin D and Brain Health

Researchers have linked hormonelike vitamin D to several illnesses across the body, including diabetes, cancer, stroke, and cardiovascular disease, with evidence from numerous studies [74]. The list of conditions that are associated with lower vitamin D levels now includes cognitive impairment, dementia, psychosis, and autism [75]. The central nervous system is one of many body parts containing calcium-binding proteins

[76]. Calbindin-D28K, calretinin, and parvalbumin have all been shown to be modulated by vitamin D in brain tissues or cells [77]. Both the adult and fetal brains contain large amounts of these calcium-binding proteins. Each of these proteins has a distinct distribution in the adult brain and shows temporal patterns as the brain develops and ages [78]. In several noteworthy studies, human neuroblastoma cells treated with calcitriol or its analogs over an extended period presented increased neurite outgrowth, as did those treated with other markers of differentiation [79].

Pregnancy and Fetus: Vitamin D and Brain Development

Vitamin D deficiency in the mother has a grievous effect on the fetal brain, as the fetus is utterly dependent on the maternal supply of vitamins through the placenta [80, 81]. Vitamin D plays an essential role in the healthy development of the fetal brain, as it influences both the function and structure of the developing central nervous system. Vitamin D affects the brain through its active form, calcitriol [82]. The vitamin D receptor is widely distributed throughout the human brain, particularly in the hippocampal area and in various cell types, including microglia, astrocytes, oligodendrocytes, and neurons [83–85]. Moreover, brain-derived neurotrophic factor, which is essential for neurogenesis, synaptic plasticity, and neuron survival, is also regulated by vitamin D [86]. Vitamin D plays a crucial role in the development of the cortex, hippocampus, and cerebellum and thus influences motor and cognitive function later in life [84, 87]. In earlier stages of life, vitamin D is also required for neurotransmission, as the synthesis and metabolism of different neurotransmitters, e.g., serotonin and synaptic protein expression, are influenced by vitamin D [88]. Increased levels of inflammation during gestation can disorganize the normal development of the fetal brain [89] and increase the risk of neurodevelopmental disorders [90]. The anti-inflammatory activity of vitamin D contributes to healthy neurodevelopment in fetuses [91].

Neonates and Adolescence: Vitamin D and Brain Development

Early childhood and adolescence are periods when the brain undergoes remarkable growth and development [92]. The synthesis of new synapses and myelination occur expeditiously, especially during the first 3 years of life, and remodeling of the central nervous system continues until young adulthood [93]. It has been demonstrated that maternal deficiency in vitamin D may lead to disturbances in neurodevelopment in neonates [94]. Insufficient levels of vitamin D may contribute to developmental delay and increase the risk of attention deficit hyperactive disorder and autism spectrum disorders [95–97]. However, the mechanism is not precise [98]. The neuroprotective ability of vitamin D reduces oxidative stress in neonates with hypoxic-ischemic encephalopathy and thus results in positive developmental effects [99]. The scarcity of vitamin D during the antenatal period may lead to a reduction in white and gray matter volume [100] and induce behavioral and cognitive disturbances along with psychotic disorders [101]. Vitamin D-deficient children also exhibit lower intelligence quotient [102] and less social competency [103].

Vitamin D Deficiency-Related Neurological Diseases and Management Strategies

The evidence regarding the role of vitamin D deficiency in several neurological diseases, along with possible therapies, is discussed in this section. The association between 25-hydroxycholecalciferol levels and the risk of developing Parkinson's disease remains unclear [104]. Knekt et al. [105] conducted a study in Finland involving 3173 recruits over 29 years and observed 50 individuals who developed Parkinson's disease [105]. A previous study reported that low vitamin D levels were of predictive value for increased neurodegenerative disease incidence [106]. However, Fullard et al. [107] reported no link between vitamin D deficiency and Parkinson's disease pathogenesis [107]. Another study by Shrestha et al. [108]

agreed with the survey outcome of Fullard et al. [107]. This difference in findings may be due to differences in the race and geographical location of the study participants. Wang et al. [109] reported that single-nucleotide polymorphisms in the vitamin D receptor (FokI) are related to a lower risk of Parkinson's disease in Asian populations than in Caucasian populations [109]. Vitamin D receptor polymorphisms are possibly responsible for Parkinson's disease development [110, 111]. Several studies have revealed that low 25-hydroxycholecalciferol has an inverse association with the severity of motor symptoms in Parkinson's disease patients [106, 112–114]. Muscle strength is influenced by vitamin D, and there is an increase in fracture risk, which has been linked to low vitamin D levels [115–117]. Supplementation with vitamin D may positively impact patients with Parkinson's disease by reducing neuroinflammation and neural damage, especially when it is administered orally at a dosage of 800 IU/day for 12 months [118–120]. Another study reported a significant, inverse correlation between vitamin D levels and the total Unified Parkinson's Disease Rating Scale ($p = 0.004$) and the Unified Parkinson's Disease Rating Scale III ($p = 0.002$) [121]. This finding was not confirmed by other studies [122, 123]. Vitamin D supplementation remains a considerable choice of medication for Parkinson's disease patients, as it is associated with few risk factors and offers substantial benefits [124, 125]. Further clinical trials are needed to support this therapeutic option.

Alzheimer's disease is a neurodegenerative dementia that is characterized by the deposition of extracellular amyloid beta, pathological intracellular tangles of tau protein, and loss of neurons [126–128]. Several studies have revealed that individuals with low 25-hydroxycholecalciferol levels in the serum are more susceptible to the development of vascular dementia and Alzheimer's disease [129–137], especially in elderly communities, in terms of cognitive function and dementia [138, 139], possibly because of a diminished hippocampal mass and reduced neuropsychological performance [140]. Extreme vitamin D paucity (<10 ng/ml) and dementia and Alzheimer's disease equated to tolerable vitamin D insufficiency (10–20 ng/ml) [141]. Another study reported that in animal models, vitamin D insufficiency exacerbates Alzheimer-like pathologies by increasing amyloid-beta (Aβ) generation, inducing inflammatory stress, and increasing Tau phosphorylation through a decrease in the antioxidant capacity of the brains of human amyloid precursor protein/presenilin 1 [142]. However, other studies have not confirmed this association, predominantly in middle-aged and elderly individuals, regarding cognitive function and dementia [130, 143–152]. 1,25-Dihydroxycholecalciferol has been found to improve the phagocytosis of amyloid plaques and immune cells, particularly macrophage clearance [153, 154]. Vitamin D is also involved in modulating the response to inflammation, mitochondrial respiratory function, and oxidative stress (Fig. 1), which affects the pathogenesis of Alzheimer's disease [155–161]. A decrease in serum vitamin D that promotes Alzheimer's disease has been noted in individuals with genetically increased levels of 25-hydroxycholecalciferol [162, 163]. However, multiple studies have shown the benefit of vitamin D supplementation among Alzheimer's patients, with the goal of achieving a serum level of >30 ng/mL [164–169]. Another earlier report suggested that supplementation with 800–1000 IU/day of vitamin D or 50,000 IU every 30 days is considered protective for the majority of patients and supports the desired therapeutic range of vitamin D [170]. However, Aspell et al. [171] reported that although epidemiologic data support the associations between low serum 25-hydroxycholecalciferol concentrations and poorer cognitive performance among older people with Alzheimer's disease living in the community, an ideal 25-hydroxycholecalciferol level for cognitive health could not be determined [171]. Nonetheless, no evidence suggests that vitamin D supplementation negatively affects patients with Alzheimer's disease [172]; nevertheless, prolonged supplementation is not advocated [173, 174]. It has been reported that the Mini-

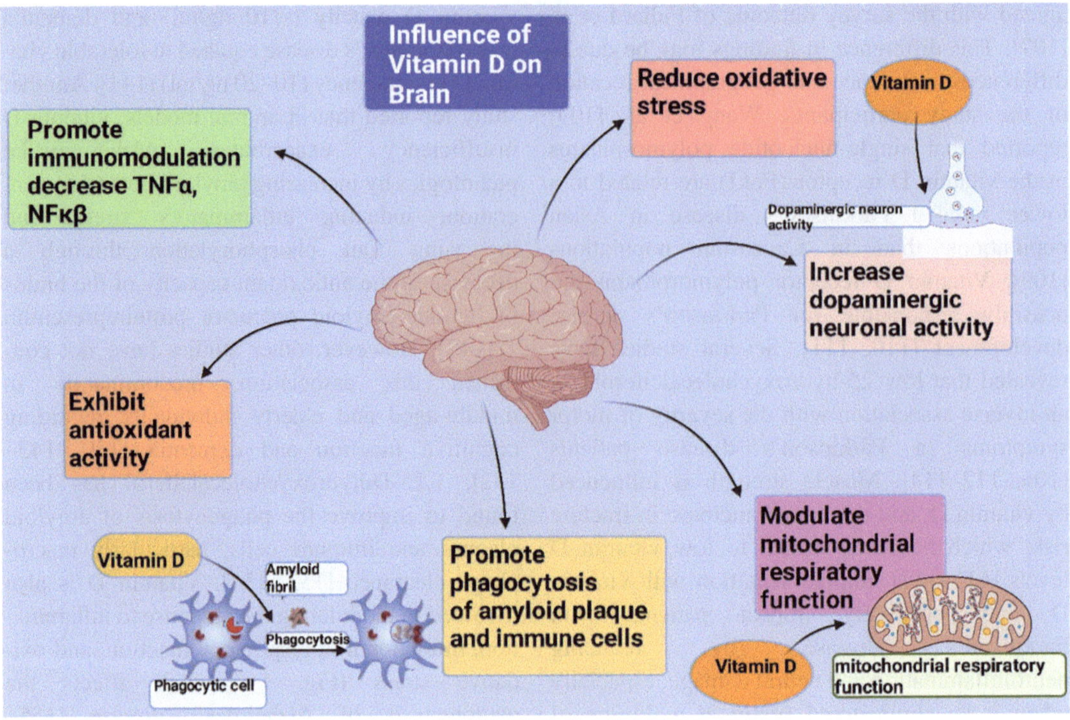

Fig. 1 Role of vitamin D in brain health and inflammation. NF-κB: nuclear factor kappa beta; TNFα: tumor necrosis factor-α. This figure was drawn via the premium version of BioRender (https://biorender.com/), accessed on January 12, 2025, with license number DG27QJLS5K. (Illustration Credit: Rahnuma Ahmad)

Mental State Examination score and serum 25-hydroxycholecalciferol concentration may be beneficial biomarkers for the estimation and identification of mild cognitive impairment (and various stages of Alzheimer's disease) [175]. Bivona et al. [176] revealed that vitamin D is not a reliable biomarker of Alzheimer's disease, as measuring this biomarker does not improve the diagnosis and prognosis of these patients [176]. Thus, more research with clinical trials is needed to determine whether vitamin D may aid in halting the progression of this neurodegenerative condition.

Vitamin D strongly influences the pathogenesis of multiple sclerosis (demyelinating disease of the central nervous system). Several studies have reported an increasing incidence of multiple sclerosis with increasing latitude, which indicates less sunlight exposure [177–182]. The frequency of multiple sclerosis has been reported to decrease even at high latitudes when fatty fish rich in vitamin D are consumed [183, 184]. The use of sunscreen, which mainly contains salates, minimizes the risk of multiple sclerosis [185]. 25-Hydroxycholecalciferol levels are lower in individuals who develop multiple sclerosis; thus, low vitamin D levels may be one of the risk factors for multiple sclerosis [186, 187]. Low levels of vitamin D at the beginning of multiple sclerosis may lead to early conversion to secondary progression [161, 162]. Several studies have reported that low vitamin D levels are associated with the progression of multiple sclerosis and clinical disability [188–195].

In the United States, a clinical trial involving 172 individuals with multiple sclerosis, who were divided into 2 groups, was conducted between 2012 and 2019. One group received 600 IU/day (low dose) of vitamin D, and the other received 5000 IU/day (high dose) of vitamin D in addition to glatiramer acetate. However, no significant difference was noted in the proportion of recruits suffering relapse and those experiencing relapse being annulled [196]. On the other hand, two other randomized controlled trials reported a positive influence of vitamin D supplementation, with a lower number of new or enlarging lesions at T2, a reduced lesion volume at T1, and a slower progression of disability [188, 197, 198]. Vitamin D supplementation may be considered to prevent low vitamin levels in those with multiple sclerosis, and levels of approximately 100 nmol/L and higher should be maintained [199].

Several cross-sectional and case-control studies have reported associations between headaches and low vitamin D levels in the serum [200–211]. Patients with tension-type headaches and migraine have been reported to have lower 25-hydroxycholecalciferol serum levels, but whether a headache may cause vitamin D deficiency or whether vitamin D deficiency can result in headache has yet to be determined [202, 212, 213]. In various studies, supplementation with vitamin D at varying dosages and for multiple periods of follow-up has been shown to significantly decrease the number of headache attacks and the frequency of headache episodes. However, the reduction in pain intensity was less evident [214–217].

Toxicities of Vitamin D Overconsumption and Brain Health

Vitamin D toxicity ensues because of the consumption of supplements redundantly, incorrect prescriptions, or, less commonly, contaminated food, and its incidence has been on the rise for the last decade [218, 219]. Overdosing causes increased levels of calcium ions in the blood and inequality in the organization of bone metabolism, which in turn induces toxicity [220]. According to a report from America's Poison Centers, the number of vitamin D toxicity cases has surpassed the total number of other vitamin toxicity cases combined [221]. Initial symptoms are nonspecific, but severe symptoms manifest as neurological features such as apathy, confusion, agitation, ataxia, and stupor and may even lead to coma [58, 222, 223]. Central pontine myelinolysis due to chronic VD overdose has been reported [224]. Several animal studies have demonstrated that extremely high serum levels of vitamin D may induce wave changes in electroencephalograms (EEGs) [225], premature aging [226], and impaired cognitive function [227].

Conclusion

This chapter discusses in detail the role of vitamin D in brain health. This study provides information regarding the physiological functions of vitamin D, its role in pregnancy and fetal development, its effects on neurological disease development, its role as a supplement in such disease conditions, and the possible toxicity of vitamin D. Research has shown its effect as a neuroprotective agent through reducing pro-inflammatory cytokines, promoting mitochondrial health, lowering oxidative stress, and forming antioxidants. Studies have also noted the role of vitamin D in brain development in fetuses and during the growth stage of children. The role of vitamin D in neuronal health is partially understood. Therefore, further studies are needed to understand the mechanism of action of vitamin D in the central nervous system and the benefits of vitamin D supplementation. Extensive studies may also help reveal the advantages and disadvantages, if any, of using vitamin D in the general population as a regular supplement. Therefore, it can be suggested that the role of this sunshine vitamin is not limited to calcium ion and bone homeostasis but extends to immune and neuronal health maintenance. For updated guidelines on vitamin D testing and dosing, readers are referred to a recent publication [228] and Chap. 2 of this book [229].

References

1. McCollum EV, Simmonds BJE et al (1922) Studies on experimental rickets: xxi. An experimental demonstration of the existence of a vitamin which

promotes calcium deposition. J Biol Chem 53(2): 293–312. https://doi.org/10.1016/S0021-9258(18)85783-0
2. Holick MF (2023) The one-hundred-year anniversary of the discovery of the sunshine vitamin D_3: historical, personal experience and evidence-based perspectives. Nutrients 15(3):593. https://doi.org/10.3390/nu15030593
3. Hernigou P, Auregan JC, Dubory A (2019) Vitamin D: part II; cod liver oil, ultraviolet radiation, and eradication of rickets. Int Orthop 43(3):735–749. https://doi.org/10.1007/s00264-019-04288-z
4. Dahash BA, Sankararaman S (2025) Rickets. [Updated 2023 Aug 7]. In: StatPearls [Internet]. StatPearls Publishing, Treasure Island. Available from: https://www.ncbi.nlm.nih.gov/books/NBK562285/. Accessed 23 May 2025
5. Gharibzahedi SMT, Moghadam M, Amft J et al (2023) Recent advances in dietary sources, health benefits, emerging encapsulation methods, food fortification, and new sensor-based monitoring of vitamin B12: a critical review. Molecules 28(22):7469. https://doi.org/10.3390/molecules28227469
6. Polzonetti V, Pucciarelli S, Vincenzetti S et al (2020) Dietary intake of vitamin D from dairy products reduces the risk of osteoporosis. Nutrients 12(6): 1743. https://doi.org/10.3390/nu12061743
7. Meitern R, Valge M et al (2025) Birth season associates with multiple anthropometric traits in Estonian children. J Dev Orig Health Dis 16:e19. https://doi.org/10.1017/S2040174425000108
8. Holick MF (2020) Sunlight, UV radiation, vitamin D, and skin cancer: how much sunlight do we need? Adv Exp Med Biol 1268:19–36. https://doi.org/10.1007/978-3-030-46227-7_2
9. Chang SW, Lee HC (2019) Vitamin D and health — the missing vitamin in humans. Pediatr Neonatol 60(3):237–244. https://doi.org/10.1016/j.pedneo.2019.04.007
10. Ashley RD (2018) Ask the doctors – how much sunshine do I need for enough vitamin D? UCLA Health. Available at https://www.uclahealth.org/news/article/ask-the-doctors-round-sun-exposure-vital-to-vitamin-d-production. Accessed 17 Dec 2024
11. Carmeliet G, Dermauw V, Bouillon R (2015) Vitamin D signaling in calcium and bone homeostasis: a delicate balance. Best Pract Res Clin Endocrinol Metab 29(4):621–631. https://doi.org/10.1016/j.beem.2015.06.001
12. Bhattarai HK, Shrestha S, Rokka K et al (2020) Vitamin D, calcium, parathyroid hormone, and sex steroids in bone health and effects of aging. J Osteoporos 2020:9324505. https://doi.org/10.1155/2020/9324505
13. Szymczak I, Pawliczak R (2016) The active metabolite of vitamin D3 as a potential Immunomodulator. Scand J Immunol 83(2):83–91. https://doi.org/10.1111/sji.12403
14. Szymczak-Pajor I, Drzewoski J, Śliwińska A (2020) The molecular mechanisms by which vitamin D prevents insulin resistance and associated disorders. Int J Mol Sci 21(18):6644. https://doi.org/10.3390/ijms21186644
15. Fleet JC (2022) Vitamin D-mediated regulation of intestinal calcium absorption. Nutrients 14(16): 3351. https://doi.org/10.3390/nu14163351
16. Lu X, Chen Z, Vick S et al (2019) Vitamin D receptor and metabolite effects on corneal epithelial cell gap junction proteins. Exp Eye Res 187:107776. https://doi.org/10.1016/j.exer.2019.107776
17. AlGhamdi SA, Enaibsi NN, Alsufiani HM et al (2022) A single oral vitamin D_3 bolus reduces inflammatory markers in healthy Saudi males. Int J Mol Sci 23(19):11992. https://doi.org/10.3390/ijms231911992
18. Gorimanipalli B, Shetty R, Sethu S et al (2023) Vitamin D and eye: current evidence and practice guidelines. Indian J Ophthalmol 71(4):1127–1134. https://doi.org/10.4103/IJO.IJO_3174_22
19. Rolando M, Barabino S (2023) Dry eye disease: what is the role of vitamin D? Int J Mol Sci 24(2):1458. https://doi.org/10.3390/ijms24021458
20. Grimm MOW, Thiel A, Lauer AA et al (2017) Vitamin D and its analogues decrease amyloid-β (Aβ) formation and increase Aβ-degradation. Int J Mol Sci 18(12):2764. https://doi.org/10.3390/ijms18122764
21. Mrugacz M, Pieńczykowska K, Bryl A (2024) The role of vitamin D3 in ocular diseases. Nutrients 16(12):1878. https://doi.org/10.3390/nu16121878
22. de la Guía-Galipienso F, Martínez-Ferran M, Vallecillo N et al (2021) Vitamin D and cardiovascular health. Clin Nutr 40(5):2946–2957. https://doi.org/10.1016/j.clnu.2020.12.025
23. Haider F, Ghafoor H, Hassan OF et al (2023) Vitamin D and cardiovascular diseases: an update. Cureus 15(11):e49734. https://doi.org/10.7759/cureus.49734
24. Hazique M, Khan KI, Ramesh P et al (2022) A study of vitamin D and its correlation with severity and complication of congestive heart failure: a systematic review. Cureus 14(9):e28873. https://doi.org/10.7759/cureus.28873
25. Cosentino N, Campodonico J, Milazzo V et al (2021) Vitamin D and cardiovascular disease: current evidence and future perspectives. Nutrients 13(10): 3603. https://doi.org/10.3390/nu13103603
26. Razzaque MS (2018) Can adverse effects of excessive vitamin D supplementation occur without developing hypervitaminosis D? J Steroid Biochem Mol Biol 180:81–86. https://doi.org/10.1016/j.jsbmb.2017.07.006
27. Izzo M, Carrizzo A, Izzo C et al (2021) Vitamin D: not just bone metabolism but a key player in cardiovascular diseases. Life (Basel) 11(5):452. https://doi.org/10.3390/life11050452
28. Secondulfo C, Visco V, Virtuoso N et al (2024) Vitamin D: a bridge between kidney and heart. Life (Basel) 14(5):617. https://doi.org/10.3390/life14050617
29. Trehan N, Afonso L, Levine DL et al (2017) Vitamin D deficiency, supplementation, and cardiovascular

health. Crit Pathw Cardiol 16(3):109–118. https://doi.org/10.1097/HPC.0000000000000122
30. Pella D, Poruban T, Gonsorcik J et al (2024) Assessment of vitamin D levels in correlation with coronary computed tomography angiography results in a Slovak population. Arch Med Sci 21(2):409–415. https://doi.org/10.5114/aoms/188717
31. Carlberg C (2022) Vitamin D and its target genes. Nutrients 14(7):1354. https://doi.org/10.3390/nu14071354
32. Bikle DD (2016) Extraskeletal actions of vitamin D. Ann N Y Acad Sci 1376(1):29–52. https://doi.org/10.1111/nyas.13219
33. Chen S, Sun Y, Agrawal DK (2015) Vitamin D deficiency and essential hypertension. J Am Soc Hypertens 9(11):885–901. https://doi.org/10.1016/j.jash.2015.08.009. Epub 2015 Aug 21
34. Immanuel J, Yun S (2023) Vascular inflammatory diseases and endothelial phenotypes. Cells 12(12): 1640. https://doi.org/10.3390/cells12121640
35. Dubourg V, Schwerdt G, Schreier B et al (2023) EGFR activation differentially affects the inflammatory profiles of female human aortic and coronary artery endothelial cells. Sci Rep 13(1):22827. https://doi.org/10.1038/s41598-023-50148-7
36. Tonami K, Hayashi T, Uchijima Y et al (2023) Coordinated linear and rotational movements of endothelial cells compartmentalized by VE-cadherin drive angiogenic sprouting. iScience 26(7):107051. https://doi.org/10.1016/j.isci.2023.107051
37. Gogiraju R, Bochenek ML, Schäfer K (2019) Angiogenic endothelial cell signaling in cardiac hypertrophy and heart failure. Front Cardiovasc Med 6:20. https://doi.org/10.3389/fcvm.2019.00020
38. Koenigsberger M, Sauser R, Bény JL et al (2005) Role of the endothelium on arterial vasomotion. Biophys J 88(6):3845–3854. https://doi.org/10.1529/biophysj.104.054965
39. Kumar G, Dey SK, Kundu S (2020) Functional implications of vascular endothelium in regulation of endothelial nitric oxide synthesis to control blood pressure and cardiac functions. Life Sci 259:118377. https://doi.org/10.1016/j.lfs.2020.118377
40. Pi X, Xie L, Patterson C (2018) Emerging roles of vascular endothelium in metabolic homeostasis. Circ Res 123(4):477–494. https://doi.org/10.1161/CIRCRESAHA.118.313237
41. Cimmino G, Conte S, Morello M et al (2022) Vitamin D inhibits IL-6 pro-Atherothrombotic effects in human endothelial cells: a potential mechanism for protection against COVID-19 infection? J Cardiovasc Dev Dis 9(1):27. https://doi.org/10.3390/jcdd9010027
42. Martinez-Moreno JM, Herencia C, Montes de Oca A et al (2016) Vitamin D modulates tissue factor and protease-activated receptor 2 expression in vascular smooth muscle cells. FASEB J 30(3):1367–1376. https://doi.org/10.1096/fj.15-272872
43. Mohammad S, Mishra A, Ashraf MZ (2019) Emerging role of vitamin D and its associated molecules in pathways related to pathogenesis of thrombosis. Biomolecules 9(11):649. https://doi.org/10.3390/biom9110649
44. Sun D, Luo F, Xing JC, Zhang F, Xu JZ, Zhang ZH (2018) 1,25(OH)$_2$ D$_3$ inhibited Th17 cells differentiation via regulating the NF-κB activity and expression of IL-17. Cell Prolif 51(5):e12461. https://doi.org/10.1111/cpr.12461
45. Dusso AS, Bauerle KT, Bernal-Mizrachi C (2021) Non-classical vitamin D actions for renal protection. Front Med (Lausanne) 8:790513. https://doi.org/10.3389/fmed.2021.790513
46. Gonçalves LED, Andrade-Silva M, Basso PJ et al (2023) Vitamin D and chronic kidney disease: insights on lipid metabolism of tubular epithelial cell and macrophages in tubulointerstitial fibrosis. Front Physiol 14:1145233. https://doi.org/10.3389/fphys.2023.1145233
47. Cordero JB, Cozzolino M, Lu Y et al (2002) 1,25-Dihydroxyvitamin D down-regulates cell membrane growth- and nuclear growth-promoting signals by the epidermal growth factor receptor. J Biol Chem 277(41):38965–38971. https://doi.org/10.1074/jbc.M203736200
48. Wu EL, Cui HX (2016) Effect of 1,25-(OH)2D3 and lipopolysaccharide on mononuclear cell inflammation in type 2 diabetes mellitus and diabetic nephropathy uremia. Genet Mol Res 15(3). https://doi.org/10.4238/gmr.15038553
49. Yang S, Li A, Wang J et al (2018) Vitamin D receptor: a novel therapeutic target for kidney diseases. Curr Med Chem 25(27):3256–3271. https://doi.org/10.2174/0929867325666180214122352
50. Golmohammadi MG, Banaei S, Nejati K et al (2020) Vitamin D3 and erythropoietin protect against renal ischemia-reperfusion injury via heat shock protein 70 and microRNA-21 expression. Sci Rep 10(1): 20906. https://doi.org/10.1038/s41598-020-78045-3
51. Arfian N, Muflikhah K, Soeyono SK et al (2016) Vitamin D attenuates kidney fibrosis via reducing fibroblast expansion, inflammation, and epithelial cell apoptosis. Kobe J Med Sci 62(2):E38–E44. https://pmc.ncbi.nlm.nih.gov/articles/PMC5425134/pdf/e38.pdf
52. Wu CC, Liao MT, Hsiao PJ et al (2020) Antiproteinuria effect of calcitriol in patients with chronic kidney disease and vitamin D deficiency: a randomized controlled study. J Ren Nutr 30(3): 200–207. https://doi.org/10.1053/j.jrn.2019.09.001
53. Restrepo Valencia CA, Aguirre Arango JV (2016) Vitamin D (25(OH)D) in patients with chronic kidney disease stages 2-5. Colomb Med (Cali) 47(3): 160–166. Available at https://pmc.ncbi.nlm.nih.gov/articles/PMC5091275/pdf/1657-9534-cm-47-03-00160.pdf. Accessed 23 May 2025
54. Giannini S, Mazzaferro S, Minisola S et al (2017) Raising awareness on the therapeutic role of cholecalciferol in CKD: a multidisciplinary-based opinion. Endocrine 59(2):242–259. https://doi.org/10.1007/s12020-017-1369-3
55. De Jong MA, Eisenga MF et al (2021) Fibroblast growth factor 23 and new-onset chronic kidney

disease in the general population: the prevention of renal and vascular Endstage disease (PREVEND) study. Nephrol Dial Transplant 36(1):121–128. https://doi.org/10.1093/ndt/gfz266
56. Latic N, Erben RG (2021) FGF23 and vitamin D metabolism. JBMR Plus 5(12):e10558. https://doi.org/10.1002/jbm4.10558
57. Fakhoury M, Levy R, Melamed ML (2019) Vitamin D deficiency and kidney hyperfiltration: a mechanism of kidney injury? Ann Transl Med 7(Suppl 6):S207. https://doi.org/10.21037/atm.2019.06.64
58. Marcinowska-Suchowierska E, Kupisz-Urbańska M, Łukaszkiewicz J et al (2018) Vitamin D toxicity-a clinical perspective. Front Endocrinol (Lausanne) 9: 550. https://doi.org/10.3389/fendo.2018.00550
59. Ray M, Jovanovich A (2019) Mineral bone abnormalities and vascular calcifications. Adv Chronic Kidney Dis 26(6):409–416. https://doi.org/10.1053/j.ackd.2019.09.004
60. Wang J, Zhou JJ, Robertson GR et al (2018) Vitamin D in vascular calcification: a double-edged sword? Nutrients 10(5):652. https://doi.org/10.3390/nu10050652
61. Pérez-Barrios C, Hernández-Álvarez E, Blanco-Navarro I et al (2016) Prevalence of hypercalcemia related to hypervitaminosis D in clinical practice. Clin Nutr 35(6):1354–1358. https://doi.org/10.1016/j.clnu.2016.02.017
62. Mason N, Sweet LM, Wills Z et al (2024) Vitamin D abuse in pursuit of "gains" resulting in acute kidney injury. Mil Med 189(1–2):e417–e419. https://doi.org/10.1093/milmed/usad237
63. Nair S (2010) Vitamin D deficiency and liver disease. Gastroenterol Hepatol 6(8):491–493. https://pmc.ncbi.nlm.nih.gov/articles/PMC2950664/pdf/GH-06-491.pdf
64. Kitson MT, Roberts SK (2012) D-livering the message: the importance of vitamin D status in chronic liver disease. J Hepatol 57(4):897–909. https://doi.org/10.1016/j.jhep.2012.04.033
65. Hoan NX, Khuyen N, Binh MT et al (2016) Association of vitamin D deficiency with hepatitis B virus-related liver diseases. BMC Infect Dis 16(1):507. https://doi.org/10.1186/s12879-016-1836-0
66. Zhao XY, Li J, Wang JH et al (2016) Vitamin D serum level is associated with child-Pugh score and metabolic enzyme imbalances but not viral load in chronic hepatitis B patients. Medicine (Baltimore) 95(27):e3926. https://doi.org/10.1097/MD.0000000000003926
67. Petta S, Cammà C, Scazzone C et al (2010) Low vitamin D serum level is related to severe fibrosis and low responsiveness to interferon-based therapy in genotype 1 chronic hepatitis C. Hepatology 51(4): 1158–1167. https://doi.org/10.1002/hep.23489
68. Terrier B, Carrat F, Geri G et al (2011) Low 25-OH vitamin D serum levels correlate with severe fibrosis in HIV-HCV co-infected patients with chronic hepatitis. J Hepatol 55(4):756–761. https://doi.org/10.1016/j.jhep.2011.01.041
69. Barchetta I, Cimini FA, Cavallo MG (2017) Vitamin D supplementation and nonalcoholic fatty liver disease: present and future. Nutrients 9(9):1015. https://doi.org/10.3390/nu9091015
70. Hoan NX, Tong HV, Song LH et al (2018) Vitamin D deficiency and hepatitis virus-associated liver diseases: a literature review. World J Gastroenterol 24(4):445–460. https://doi.org/10.3748/wjg.v24.i4.445
71. Kumar M, Parchani A, Kant R et al (2023) Relationship between vitamin D deficiency and nonalcoholic fatty liver disease: a cross-sectional study from a tertiary care center in Northern India. Cureus 15(2): e34921. https://doi.org/10.7759/cureus.34921
72. Eliades M, Spyrou E (2015) Vitamin D: a new player in nonalcoholic fatty liver disease? World J Gastroenterol 21(6):1718–1727. https://doi.org/10.3748/wjg.v21.i6.1718
73. Lee SB, Jin MH, Yoon JH (2024) The contribution of vitamin D insufficiency to the onset of steatotic liver disease among individuals with metabolic dysfunction. Sci Rep 14(1):6714. https://doi.org/10.1038/s41598-024-57380-9
74. Baneen U, Naseem S (2019) Correlation of severity of chronic obstructive pulmonary disease with serum vitamin-D level. J Family Med Prim Care 8(7): 2268–2277. https://doi.org/10.4103/jfmpc.jfmpc_404_19
75. Martin-Gorgojo A, Martin-Moreno JM (2024) Insights into the role of vitamin D in the prevention and control of cancer and other chronic noncommunicable diseases: shedding further light on a captivating subject. Nutrients 16(13):2166. https://doi.org/10.3390/nu16132166
76. Lerner PP, Sharony L, Miodownik C (2018) Association between mental disorders, cognitive disturbances, and vitamin D serum level: current state. Clin Nutr ESPEN 23:89–102. https://doi.org/10.1016/j.clnesp.2017.11.011
77. Elíes J, Yáñez M, Pereira TMC et al (2020) An update to calcium binding proteins. Adv Exp Med Biol 1131:183–213. https://doi.org/10.1007/978-3-030-12457-1_8
78. Acharya M, Singh N, Gupta G et al (2024) Vitamin D, Calbindin, and calcium signaling: unraveling the Alzheimer's connection. Cell Signal 116:111043. https://doi.org/10.1016/j.cellsig.2024.111043
79. Schieweck R, Ninkovic J, Kiebler MA (2021) RNA-binding proteins balance brain function in health and disease. Physiol Rev 101(3):1309–1370. https://doi.org/10.1152/physrev.00047.2019
80. Pet MA, Brouwer-Brolsma EM (2016) The impact of maternal vitamin D status on offspring brain development and function: a systematic review. Adv Nutr 7(4):665–678. https://doi.org/10.3945/an.115.010330

81. Larqué E, Morales E, Leis R et al (2018) Maternal and foetal health implications of vitamin D status during pregnancy. Ann Nutr Metab 72(3):179–192. https://doi.org/10.1159/000487370
82. Jeong SP, Sharma N, An SSA (2024) Role of calcitriol and vitamin D receptor (VDR) gene polymorphisms in Alzheimer's disease. Int J Mol Sci 25(9):4806. https://doi.org/10.3390/ijms25094806
83. Lasoń W, Jantas D, Leśkiewicz M et al (2023) The vitamin D receptor as a potential target for the treatment of age-related neurodegenerative diseases such as Alzheimer's and Parkinson's diseases: a narrative review. Cells 12(4):660. https://doi.org/10.3390/cells12040660
84. Eyles DW (2020) Vitamin D: brain and behavior. JBMR Plus 5(1):e10419. https://doi.org/10.1002/jbm4.10419
85. Lardner AL (2015) Vitamin D and hippocampal development-the story so far. Front Mol Neurosci 8: 58. https://doi.org/10.3389/fnmol.2015.00058
86. Khairy EY, Attia MM (2021) Protective effects of vitamin D on neurophysiologic alterations in brain aging: role of brain-derived neurotrophic factor (BDNF). Nutr Neurosci 24(8):650–659. https://doi.org/10.1080/1028415X.2019.1665854
87. Cui X, Gooch H, Petty A et al (2017) Vitamin D, and the brain: genomic and non-genomic actions. Mol Cell Endocrinol 453:131–143. https://doi.org/10.1016/j.mce.2017.05.035
88. Ye X, Zhou Q, Ren P et al (2023) The synaptic and circuit functions of vitamin D in neurodevelopment disorders. Neuropsychiatr Dis Treat 19:1515–1530. https://doi.org/10.2147/NDT.S407731
89. Gyllenhammer LE, Rasmussen JM, Bertele N et al (2022) Maternal inflammation during pregnancy and offspring brain development: the role of mitochondria. Biol Psychiatry Cogn Neurosci Neuroimaging 7(5):498–509. https://doi.org/10.1016/j.bpsc.2021.11.003
90. Han VX, Patel S, Jones HF et al (2021) Maternal acute and chronic inflammation in pregnancy is associated with common neurodevelopmental disorders: a systematic review. Transl Psychiatry 11(1):71. https://doi.org/10.1038/s41398-021-01198-w
91. Lisi G, Ribolsi M, Siracusano A et al (2020) Maternal vitamin D and its role in determining fetal origins of mental health. Curr Pharm Des 26(21):2497–2509. https://doi.org/10.2174/1381612826666200506093858
92. Lebel C, Deoni S (2018) The development of brain white matter microstructure. NeuroImage 182:207–218. https://doi.org/10.1016/j.neuroimage.2017.12.097
93. Jeong J, Franchett EE, Ramos de Oliveira CV et al (2021) Parenting interventions to promote early child development in the first three years of life: a global systematic review and meta-analysis. PLoS Med 18(5):e1003602. https://doi.org/10.1371/journal.pmed.1003602
94. Rodgers MD, Mead MJ, McWhorter CA et al (2023) Vitamin D and child neurodevelopment-a post hoc analysis. Nutrients 15(19):4250. https://doi.org/10.3390/nu15194250
95. Chien MC, Huang CY, Wang JH et al (2024) Effects of vitamin D in pregnancy on maternal and offspring health-related outcomes: an umbrella review of systematic review and meta-analyses. Nutr Diabetes 14(1):35. https://doi.org/10.1038/s41387-024-00296-0
96. Almalki AH, Alaqel SI, Alharbi A et al (2024) Spectrofluorimetric determination of vitamin D in the serum of autistic and healthy children using functionalized graphene quantum dots. Spectrochim Acta A Mol Biomol Spectrosc 309:123842. https://doi.org/10.1016/j.saa.2024.123842
97. Lee BK, Eyles DW, Magnusson C et al (2021) Developmental vitamin D and autism spectrum disorders: findings from the Stockholm Youth Cohort. Mol Psychiatry 26(5):1578–1588. https://doi.org/10.1038/s41380-019-0578-y
98. De Marzio M, Lasky-Su J, Chu SH et al (2024) The metabolic role of vitamin D in children's neurodevelopment: a network study. Sci Rep 14(1): 16929. https://doi.org/10.1038/s41598-024-67835-8
99. Jenkins DD, Moss HG, Brown TR et al (2021) NAC and vitamin D improve CNS and plasma oxidative stress in neonatal HIE and are associated with favorable long-term outcomes. Antioxidants (Basel). 10(9):1344. https://doi.org/10.3390/antiox10091344
100. Zou R, El Marroun H, McGrath JJ et al (2020) A prospective population-based study of gestational vitamin D status and brain morphology in preadolescents. NeuroImage 209:116514. https://doi.org/10.1016/j.neuroimage.2020.116514
101. Freedman R, Hunter SK, Law AJ et al (2022) Choline, folic acid, vitamin D, and fetal brain development in the psychosis spectrum. Schizophr Res 247: 16–25. https://doi.org/10.1016/j.schres.2021.03.008
102. Cantio E, Bilenberg N, Nørgaard SM et al (2023) Vitamin D status in pregnancy and childhood associates with intelligence quotient at age 7 years: an Odense child cohort study. Aust N Z J Psychiatry 57(7):1062–1072. https://doi.org/10.1177/00048674221116027
103. López-Vicente M, Sunyer J, Lertxundi N et al (2019) Maternal circulating vitamin D3 levels during pregnancy and behavior across childhood. Sci Rep 9(1): 14792. https://doi.org/10.1038/s41598-019-51325-3
104. Xu Y, Peng J, Zhou X, Huang Y, Zhong G, Xia Z (2025) Association of 25-hydroxyvitamin D with Parkinson's disease based on the results from the NHANES 2007 to 2018 and Mendelian randomization analysis. Sci Rep 15(1):5514. https://doi.org/10.1038/s41598-025-87120-6
105. Knekt P, Kilkkinen A, Rissanen H et al (2010) Serum vitamin D and the risk of Parkinson disease. Arch

Neurol 67(7):808–811. https://doi.org/10.1001/archneurol.2010.120
106. Barichella M, Garrì F, Caronni S et al (2022) Vitamin D status and Parkinson's disease. Brain Sci 12(6):790. https://doi.org/10.3390/brainsci12060790
107. Fullard ME, Xie SX, Marek K et al (2017) Vitamin D in the Parkinson associated risk syndrome (PARS) study. Mov Disord 32(11):1636–1640. https://doi.org/10.1002/mds.27127
108. Shrestha S, Lutsey PL, Alonso A et al (2016) Serum 25-hydroxyvitamin D concentrations in mid-adulthood and Parkinson's disease risk. Mov Disord 31(7):972–978. https://doi.org/10.1002/mds.26573
109. Wang X, Shen N, Lu Y et al (2019) Vitamin D receptor polymorphisms and the susceptibility of Parkinson's disease. Neurosci Lett 699:206–211. https://doi.org/10.1016/j.neulet.2019.02.018
110. Kundu NC, Kundu A, Khalil MI et al (2025) A case-control study on vitamin D receptor gene polymorphisms in patients with Parkinson's disease in Bangladesh. Sci Rep 15(1):12333. https://doi.org/10.1038/s41598-025-96195-0
111. Agúndez JAG, García-Martín E, Alonso-Navarro H et al (2022) Vitamin D receptor and binding protein gene variants in patients with essential tremor. Mol Neurobiol 59(6):3458–3466. https://doi.org/10.1007/s12035-022-02804-8
112. Soliman RH, Oraby MI, Hussein M et al Could vitamin D deficiency have an impact on motor and cognitive function in Parkinson's disease? Egypt J Neurol Psychiatry Neurosurg 55:34. Available at https://doi.org/10.1186/s41983-019-0084-9. Accessed 28 May 2025
113. Sleeman I, Aspray T, Lawson R et al (2017) The role of vitamin D in disease progression in early Parkinson's disease. J Parkinsons Dis 7(4):669–675. https://doi.org/10.3233/JPD-171122
114. Xia M, Zhou Q (2023) Correlation between 25-hydroxy-vitamin D and Parkinson's disease. IBRO Neurosci Rep 16:162–167. https://doi.org/10.1016/j.ibneur.2023.02.006
115. Książek A, Zagrodna A, Słowińska-Lisowska M (2019) Vitamin D, skeletal muscle function and athletic performance in athletes-a narrative review. Nutrients 11(8):1800. https://doi.org/10.3390/nu11081800
116. Wei FL, Li T, Gao QY et al (2022) Association between vitamin D supplementation and fall prevention. Front Endocrinol (Lausanne) 13:919839. https://doi.org/10.3389/fendo.2022.919839
117. Halfon M, Phan O, Teta D (2015) Vitamin D: a review on its effects on muscle strength, the risk of fall, and frailty. Biomed Res Int 953241. https://doi.org/10.1155/2015/953241
118. Pignolo A, Mastrilli S, Davì C et al (2022) Vitamin D and Parkinson's disease. Nutrients 14(6):1220. https://doi.org/10.3390/nu14061220
119. Calvello R, Cianciulli A, Nicolardi G et al (2017) Vitamin D treatment attenuates neuroinflammation and dopaminergic neurodegeneration in an animal model of Parkinson's disease, shifting M1 to M2 microglia responses. J Neuroimmune Pharmacol 12(2):327–339. https://doi.org/10.1007/s11481-016-9720-7
120. Jia J, Hu J, Huo X et al (2019) Effects of vitamin D supplementation on cognitive function and blood A-β-related biomarkers in older adults with Alzheimer's disease: a randomized, double-blind, placebo-controlled trial. J Neurol Neurosurg Psychiatry 90(12):1347–1352. https://doi.org/10.1136/jnnp-2018-320199
121. Suzuki M, Yoshioka M, Hashimoto M et al (2012) 25-hydroxyvitamin D, vitamin D receptor gene polymorphisms, and severity of Parkinson's disease. Mov Disord 27(2):264–271. https://doi.org/10.1002/mds.24016
122. Luthra NS, Kim S, Zhang Y et al (2018) Characterization of vitamin D supplementation and clinical outcomes in a large cohort of early Parkinson's disease. J Clin Mov Disord 5:7. https://doi.org/10.1186/s40734-018-0074-6
123. Tripkovic L, Lambert H, Hart K et al (2012) Comparison of vitamin D2 and vitamin D3 supplementation in raising serum 25-hydroxyvitamin D status: a systematic review and meta-analysis. Am J Clin Nutr 95(6):1357–1364. https://doi.org/10.3945/ajcn.111.031070
124. Azzam AY, Ghozy S, Azab MA (2022) Vitamin D and its' role in Parkinson's disease patients with SARS-CoV-2 infection. A review article. Interdiscip Neurosurg 27:101441. https://doi.org/10.1016/j.inat.2021.101441
125. Tumas V, Aureliano MJ, Rieder CRM et al (2025) Modifiable risk factors associated with the risk of developing Parkinson's disease: a critical review. Arq Neuropsiquiatr 83(3):1–10. https://doi.org/10.1055/s-0045-1805075
126. Liu E, Zhang Y, Wang JZ (2024) Updates in Alzheimer's disease: from basic research to diagnosis and therapies. Transl Neurodegener 13(1):45. https://doi.org/10.1186/s40035-024-00432-x
127. Zheng Q, Wang X (2025) Alzheimer's disease: insights into pathology, molecular mechanisms, and therapy. Protein Cell 16(2):83–120. https://doi.org/10.1093/procel/pwae026
128. No Authors (2021) 2021 Alzheimer's disease facts and figures. Alzheimers Dement 17(3):327–406. https://doi.org/10.1002/alz.12328
129. Shen L, Ji HF (2015) Vitamin D deficiency is associated with increased risk of Alzheimer's disease and dementia: evidence from meta-analysis. Nutr J 14:76. https://doi.org/10.1186/s12937-015-0063-7
130. Jayedi A, Rashidy-Pour A, Shab-Bidar S (2019) Vitamin D status and risk of dementia and Alzheimer's disease: a meta-analysis of dose-response. Nutr

Neurosci 22(11):750–759. https://doi.org/10.1080/1028415X.2018.1436639
131. Chai B, Gao F, Wu R et al (2019) Vitamin D deficiency as a risk factor for dementia and Alzheimer's disease: an updated meta-analysis. BMC Neurol 19(1):284. https://doi.org/10.1186/s12883-019-1500-6
132. Aguilar-Navarro SG, Mimenza-Alvarado AJ, Jiménez-Castillo GA et al (2019) Association of vitamin D with mild cognitive impairment and Alzheimer's dementia in older Mexican adults. Rev Investig Clin 71(6):381–386. https://doi.org/10.24875/RIC.19003079
133. Littlejohns TJ, Henley WE, Lang IA et al (2014) Vitamin D and the risk of dementia and Alzheimer disease. Neurology 83(10):920–928. https://doi.org/10.1212/WNL.0000000000000755
134. Licher S, de Bruijn RFAG, Wolters FJ et al (2017) Vitamin D and the risk of dementia: the Rotterdam study. J Alzheimer's Dis 60(3):989–997. https://doi.org/10.3233/JAD-170407
135. Afzal S, Bojesen SE, Nordestgaard BG (2014) Reduced 25-hydroxyvitamin D and risk of Alzheimer's disease and vascular dementia. Alzheimers Dement 10(3):296–302. https://doi.org/10.1016/j.jalz.2013.05.1765
136. Aghajafari F, Pond D, Catzikiris N et al (2018) Quality assessment of systematic reviews of vitamin D, cognition, and dementia. BJPsych Open 4(4):238–249. https://doi.org/10.1192/bjo.2018.32
137. Knekt P, Sääksjärvi K, Järvinen R et al (2014) Serum 25-hydroxyvitamin D concentration and risk of dementia. Epidemiology 25(6):799–804. https://doi.org/10.1097/EDE.0000000000000175
138. Imerbsin N, Shantavasinkul PC, Witoonpanich P et al (2025) Vitamin D and cognitive impairment. Nutrients 17(8):1301. https://doi.org/10.3390/nu17081301
139. Feart C, Helmer C, Merle B et al (2017) Associations of lower vitamin D concentrations with cognitive decline and long-term risk of dementia and Alzheimer's disease in older adults. Alzheimers Dement 13(11):1207–1216. https://doi.org/10.1016/j.jalz.2017.03.003
140. Karakis I, Pase MP, Beiser A et al (2016) Association of serum vitamin D with the risk of incident dementia and subclinical indices of brain aging: the Framingham heart study. J Alzheimer's Dis 51(2):451–461. https://doi.org/10.3233/JAD-150991
141. Ulstein I, Bøhmer T (2017) Normal vitamin levels and nutritional indices in Alzheimer's disease patients with mild cognitive impairment or dementia with normal body mass indexes. J Alzheimer's Dis 55(2):717–725. https://doi.org/10.3233/JAD-160393
142. Fan YG, Pang ZQ, Wu TY et al (2020) Vitamin D deficiency exacerbates Alzheimer-like pathologies by reducing antioxidant capacity. Free Radic Biol Med 161:139–149. https://doi.org/10.1016/j.freeradbiomed.2020.10.007
143. Yang K, Chen J, Li X et al (2019) Vitamin D concentration and risk of Alzheimer disease: a meta-analysis of prospective cohort studies. Medicine (Baltimore) 98(35):e16804. https://doi.org/10.1097/MD.0000000000016804
144. Olsson E, Byberg L, Karlström B et al (2017) Vitamin D is not associated with incident dementia or cognitive impairment: an 18-y follow-up study in community-living old men. Am J Clin Nutr 105(4):936–943. https://doi.org/10.3945/ajcn.116.141531
145. Rebelos E, Tentolouris N, Jude E (2023) The role of vitamin D in health and disease: a narrative review on the mechanisms linking vitamin D with disease and the effects of supplementation. Drugs 83(8):665–685. https://doi.org/10.1007/s40265-023-01875-8
146. Rutjes AW, Denton DA, Di Nisio M et al (2018) Vitamin and mineral supplementation for maintaining cognitive function in cognitively healthy people in mid and late life. Cochrane Database Syst Rev 12(12):CD011906. https://doi.org/10.1002/14651858.CD011906.pub2
147. Jorde R, Kubiak J, Svartberg J et al (2019) Vitamin D supplementation has no effect on cognitive performance after four months in mid-aged and older subjects. J Neurol Sci 396:165–171. https://doi.org/10.1016/j.jns.2018.11.020
148. Bischoff-Ferrari HA, Vellas B, Rizzoli R et al (2020) Effect of vitamin D supplementation, omega-3 fatty acid supplementation, or a strength-training exercise program on clinical outcomes in older adults: the DO-HEALTH randomized clinical trial. JAMA 324(18):1855–1868. https://doi.org/10.1001/jama.2020.16909
149. Moran C, Scotto di Palumbo A, Bramham J et al (2018) Effects of a six-month multi-ingredient nutrition supplement intervention of Omega-3 polyunsaturated fatty acids, vitamin D, resveratrol, and whey protein on cognitive function in older adults: a randomised, double-blind, controlled trial. J Prev Alzheimers Dis 5(3):175–183. https://doi.org/10.14283/jpad.2018.11
150. Gil Martínez V, Avedillo Salas A, Santander Ballestín S (2022) Vitamin supplementation and dementia: a systematic review. Nutrients 14(5):1033. https://doi.org/10.3390/nu14051033
151. Chakkera M, Ravi N, Ramaraju R et al (2022) The efficacy of vitamin D supplementation in patients with Alzheimer's disease in preventing cognitive decline: a systematic review. Cureus 14(11):e31710. https://doi.org/10.7759/cureus.31710
152. Du Y, Liang F, Zhang L et al (2020) Vitamin D supplement for prevention of Alzheimer's disease: a systematic review and meta-analysis. Am J Ther 28(6):e638–e648. https://doi.org/10.1097/MJT.0000000000001302
153. Patel P, Shah J (2017) Role of vitamin D in amyloid clearance via LRP-1 upregulation in Alzheimer's disease: a potential therapeutic target? J Chem

Neuroanat 85:36–42. https://doi.org/10.1016/j.jchemneu.2017.06.007
154. Banerjee A, Khemka VK, Ganguly A et al (2015) Vitamin D and Alzheimer's disease: neurocognition to therapeutics. Int J Alzheimers Dis 192747. https://doi.org/10.1155/2015/192747
155. van Olst L, Simonton B, Edwards AJ et al (2025) Microglial mechanisms drive amyloid-β clearance in immunized patients with Alzheimer's disease. Nat Med 31(5):1604–1616. https://doi.org/10.1038/s41591-025-03574-1
156. de Las HN, Martín Giménez VM, Ferder L et al (2020) Implications of oxidative stress and potential role of mitochondrial dysfunction in COVID-19: therapeutic effects of vitamin D. Antioxidants (Basel) 9(9):897. https://doi.org/10.3390/antiox9090897
157. Yin K, Agrawal DK (2014) Vitamin D and inflammatory diseases. J Inflamm Res 7:69–87. https://doi.org/10.2147/JIR.S63898
158. Matta Reddy A, Iqbal M, Chopra H et al (2022) Pivotal role of vitamin D in mitochondrial health, cardiac function, and human reproduction. EXCLI J 21:967–990. https://doi.org/10.17179/excli2022-4935
159. Misrani A, Tabassum S, Yang L (2021) Mitochondrial dysfunction and oxidative stress in Alzheimer's disease. Front Aging Neurosci 13:617588. https://doi.org/10.3389/fnagi.2021.617588
160. Firoozi A, Shadi M, Rezagholizadeh A (2025) The role of low-level laser therapy in Alzheimer's disease: a review of the potential benefits of vitamin D enhancement. Lasers Med Sci 40(1):159. https://doi.org/10.1007/s10103-025-04407-w
161. Kinney JW, Bemiller SM, Murtishaw AS et al (2018) Inflammation as a central mechanism in Alzheimer's disease. Alzheimers Dement 4:575–590. https://doi.org/10.1016/j.trci.2018.06.014
162. Wang L, Qiao Y, Zhang H et al (2020) Circulating vitamin D levels and Alzheimer's disease: a Mendelian randomization study in the IGAP and UK biobank. J Alzheimer's Dis 73(2):609–618. https://doi.org/10.3233/JAD-190713
163. Yang T, Wang H, Xiong Y et al (2020) Vitamin D supplementation improves cognitive function through reducing oxidative stress regulated by telomere length in older adults with mild cognitive impairment: a 12-month randomized controlled trial. J Alzheimer's Dis 78(4):1509–1518. https://doi.org/10.3233/JAD-200926
164. Hyppönen E, Vimaleswaran KS, Zhou A (2022) Genetic determinants of 25-Hydroxyvitamin D concentrations and their relevance to public health. Nutrients 14(20):4408. https://doi.org/10.3390/nu14204408
165. Annweiler C, Karras SN, Anagnostis P et al (2014) Vitamin D supplements: a novel therapeutic approach for Alzheimer patients. Front Pharmacol 5:6. https://doi.org/10.3389/fphar.2014.00006
166. Ghahremani M, Smith EE, Chen HY et al (2023) Vitamin D supplementation and incident dementia: effects of sex, *APOE*, and baseline cognitive status. Alzheimers Dement (Amst) 15(1):e12404. https://doi.org/10.1002/dad2.12404
167. Mizwicki MT, Menegaz D, Zhang J et al (2012) Genomic and nongenomic signaling induced by 1α,25(OH)2-vitamin D3 promotes the recovery of amyloid-β phagocytosis by Alzheimer's disease macrophages. J Alzheimer's Dis 29(1):51–62. https://doi.org/10.3233/JAD-2012-110560
168. Cashman KD, Dowling KG, Škrabáková Z et al (2016) Vitamin D deficiency in Europe: pandemic? Am J Clin Nutr 103(4):1033–1044. https://doi.org/10.3945/ajcn.115.120873
169. Sandoval KE (2025) Prevalence of reported dietary supplement use in 2019 older adult (≥ 55 years) National Alzheimer's coordinating center participants with normal cognition, mild cognitive impairment or dementia. J Hum Nutr Diet 38(2):e70049. https://doi.org/10.1111/jhn.70049
170. Kennel KA, Drake MT, Hurley DL (2010) Vitamin D deficiency in adults: when to test and how to treat. Mayo Clin Proc 85(8):752–757. quiz 757–8. https://doi.org/10.4065/mcp.2010.0138
171. Aspell N, Lawlor B, O'Sullivan M (2018) Is there a role for vitamin D in supporting cognitive function as we age? Proc Nutr Soc 77(2):124–134. https://doi.org/10.1017/S0029665117004153
172. Gombart AF, Michels AJ, Eggersdorfer M (2023) There is no evidence that vitamin D supplementation drives the progression of Alzheimer's disease. Aging Cell 22(1):e13758. https://doi.org/10.1111/acel.13758
173. Bode LE, McClester Brown M, Hawes EM (2020) Vitamin D supplementation for Extraskeletal indications in older persons. J Am Med Dir Assoc 21(2):164–171. https://doi.org/10.1016/j.jamda.2019.09.021
174. Lai RH, Hsu CC, Yu BH et al (2022) Vitamin D supplementation worsens Alzheimer's progression: animal model and human cohort studies. Aging Cell 21(8):e13670. https://doi.org/10.1111/acel.13670
175. Ouma S, Suenaga M, Bölükbaşı Hatip FF et al (2018) Serum vitamin D in patients with mild cognitive impairment and Alzheimer's disease. Brain Behav 8(3):e00936. https://doi.org/10.1002/brb3.936
176. Bivona G, Lo Sasso B, Gambino CM et al (2021) The role of vitamin D as a biomarker in Alzheimer's disease. Brain Sci 11(3):334. https://doi.org/10.3390/brainsci11030334
177. Gouider R, Souissi A, Mrabet S, Gharbi A et al (2024) Environmental factors related to multiple sclerosis progression. J Neurol Sci 464:123161. https://doi.org/10.1016/j.jns.2024.123161
178. Vukusic S, Van Bockstael V, Gosselin S et al (2007) Regional variations in the prevalence of multiple sclerosis in French farmers. J Neurol Neurosurg

Psychiatry 78(7):707–709. https://doi.org/10.1136/jnnp.2006.101196
179. Hammond SR, McLeod JG, Millingen KS et al (1988) The epidemiology of multiple sclerosis in three Australian cities: Perth, Newcastle, and Hobart. Brain 111(Pt 1):1–25. https://doi.org/10.1093/brain/111.1.1
180. Miller DH, Hammond SR, McLeod JG et al (1990) Multiple sclerosis in Australia and New Zealand: are the determinants genetic or environmental? J Neurol Neurosurg Psychiatry 53(10):903–905. https://doi.org/10.1136/jnnp.53.10.903
181. Acheson ED, Bachrach CA, Wright FM (1960) Some comments on the relationship of the distribution of multiple sclerosis to latitude, solar radiation, and other variables. Acta Psychiatr Scand Suppl 35(147):132–147. https://doi.org/10.1111/j.1600-0447.1960.tb08674.x
182. Cassard SD, Fitzgerald KC, Qian P, Emrich SA, Azevedo CJ, Goodman AD, Sugar EA, Pelletier D, Waubant E, Mowry EM (2023) High-dose vitamin D_3 supplementation in relapsing-remitting multiple sclerosis: a randomized clinical trial. EClinicalMedicine 59:101957. https://doi.org/10.1016/j.eclinm.2023.101957
183. Johansson E, Guo J, Wu J et al (2025) Impact of fish consumption on disability progression in multiple sclerosis. J Neurol Neurosurg Psychiatry 96(9): 893–899. jnnp-2024-335200. https://doi.org/10.1136/jnnp-2024-335200
184. Swank RL, Lerstad O, Strøm A et al (1952) Multiple sclerosis in rural Norway its geographic and occupational incidence in relation to nutrition. N Engl J Med 246(19):721–728. https://doi.org/10.1056/NEJM195205082461901
185. Wang Y, Marling SJ, Plum LA, DeLuca HF (2017) Salate derivatives found in sunscreens block experimental autoimmune encephalomyelitis in mice. Proc Natl Acad Sci USA 114(32):8528–8531. https://doi.org/10.1073/pnas.1703995114
186. Gombash SE, Lee PW, Sawdai E et al (2022) Vitamin D as a risk factor for multiple sclerosis: immunoregulatory or neuroprotective? Front Neurol 16(13): 796933. https://doi.org/10.3389/fneur.2022.796933
187. Stridh P, Kockum I, Huang J (2021) Seasonal variability of serum 25-hydroxyvitamin D on multiple sclerosis onset. Sci Rep 11(1):20989. https://doi.org/10.1038/s41598-021-00344-0
188. Smolders J, Torkildsen Ø, Camu W et al (2019) An update on vitamin D and disease activity in multiple sclerosis. CNS Drugs 33:1187–1199. Available at https://doi.org/10.1007/s40263-019-00674-8. Accessed 27 May 2025
189. Barzegar M, Najdaghi S, Afshari-Safavi A et al (2021) Early predictors of conversion to secondary progressive multiple sclerosis. Mult Scler Relat Disord 54:103115. https://doi.org/10.1016/j.msard.2021.103115
190. Thouvenot E, Orsini M, Daures JP et al (2015) Vitamin D is associated with degree of disability in patients with fully ambulatory relapsing-remitting multiple sclerosis. Eur J Neurol 22(3):564–569. https://doi.org/10.1111/ene.12617
191. Oliveira SR, Simão ANC, Alfieri DF et al (2017) Vitamin D deficiency is associated with disability and disease progression in multiple sclerosis patients independently of oxidative and nitrosative stress. J Neurol Sci 381:213–219. https://doi.org/10.1016/j.jns.2017.07.046
192. Bäcker-Koduah P, Bellmann-Strobl J, Scheel M et al (2020) Vitamin D and disease severity in multiple sclerosis-baseline data from the randomized controlled trial (EVIDIMS). Front Neurol 11:129. https://doi.org/10.3389/fneur.2020.00129
193. Zhang Y, Liu G, Han X et al (2016) The association of serum 25-hydroxyvitamin D levels with multiple sclerosis severity and progression in a case-control study from China. J Neuroimmunol 297:127–131. https://doi.org/10.1016/j.jneuroim.2016.05.022
194. Mandia D, Ferraro OE, Nosari G et al (2014) Environmental factors and multiple sclerosis severity: a descriptive study. Int J Environ Res Public Health 11(6):6417–6432. https://doi.org/10.3390/ijerph110606417
195. Ascherio A, Munger KL, White R et al (2014) Vitamin D as an early predictor of multiple sclerosis activity and progression. JAMA Neurol 71(3): 306–314. https://doi.org/10.1001/jamaneurol.2013.5993
196. Camu W, Lehert P, Pierrot-Deseilligny C et al (2019) Cholecalciferol in relapsing-remitting MS: a randomized clinical trial (CHOLINE). Neurol Neuroimmunol Neuroinflamm 6(5):e597. https://doi.org/10.1212/NXI.0000000000000597
197. Hupperts R, Smolders J, Vieth R et al (2019) Randomized trial of daily high-dose vitamin D_3 in patients with RRMS receiving subcutaneous interferon β-1a. Neurology 93(20):e1906–e1916. https://doi.org/10.1212/WNL.0000000000008445
198. Bettencourt A, Boleixa D, Reguengo H et al (2018) Serum 25-hydroxyvitamin D levels in multiple sclerosis patients from the north of Portugal. J Steroid Biochem Mol Biol 180:137–141. https://doi.org/10.1016/j.jsbmb.2017.09.016
199. Prakash S, Rathore C, Makwana P et al (2017) Vitamin D deficiency in patients with chronic tension-type headache: a case-control study. Headache 57(7): 1096–1108. https://doi.org/10.1111/head.13096
200. Buettner C, Burstein R (2015) Association of statin use and risk for severe headache or migraine by serum vitamin D status: a cross-sectional population-based study. Cephalalgia 35(9):757–766. https://doi.org/10.1177/0333102414559733
201. Celikbilek A, Gocmen AY, Zararsiz G et al (2014) Serum levels of vitamin D, vitamin D-binding protein and vitamin D receptor in migraine patients from

central Anatolia region. Int J Clin Pract 68(10): 1272–1277. https://doi.org/10.1111/ijcp.12456
202. Liampas I, Bourlios S, Siokas V et al (2024) Vitamin D and tension-type headache: causal association or epiphenomenon? Int J Neurosci 134(5):441–451. https://doi.org/10.1080/00207454.2022.2110495
203. Virtanen JK, Giniatullin R, Mäntyselkä P et al (2017) Low serum 25-hydroxyvitamin D is associated with higher risk of frequent headache in middle-aged and older men. Sci Rep 7:39697. https://doi.org/10.1038/srep39697
204. Petrenya N, Lamberg-Allardt C, Melhus M et al (2020) Vitamin D status in a multi-ethnic population of northern Norway: the SAMINOR 2 clinical survey. Public Health Nutr 23(7):1186–1200. https://doi.org/10.1017/S1368980018003816
205. Iannacchero R, Costa A, Squillace A et al (2015) P060. Vitamin D deficiency in episodic migraine, chronic migraine, and medication-overuse headache patients. J Headache Pain 16(Suppl 1):A184. https://doi.org/10.1186/1129-2377-16-S1-A184
206. Rapisarda L, Mazza MR, Tosto F et al (2018) Relationship between severity of migraine and vitamin D deficiency: a case-control study. Neurol Sci 39(Suppl 1):167–168. https://doi.org/10.1007/s10072-018-3384-4
207. Donmez A, Orun E, Sonmez FM (2018) Vitamin D status in children with headache: a case-control study. Clin Nutr ESPEN 23:222–227. https://doi.org/10.1016/j.clnesp.2017.09.010
208. Togha M, Razeghi Jahromi S, Ghorbani Z et al (2018) Serum vitamin D status in a group of migraine patients compared with healthy controls: a case-control study. Headache 58(10):1530–1540. https://doi.org/10.1111/head.13423
209. Song TJ, Chu MK, Sohn JH et al (2018) Effect of vitamin D deficiency on the frequency of headaches in migraine. J Clin Neurol 14(3):366–373. https://doi.org/10.3988/jcn.2018.14.3.366
210. Hancı F, Kabakuş N, Türay S et al (2020) The role of obesity and vitamin D deficiency in primary headaches in childhood. Acta Neurol Belg 120(5): 1123–1131. https://doi.org/10.1007/s13760-019-01134-2
211. Liampas I, Siokas V, Brotis A et al (2020) Vitamin D serum levels in patients with migraine: a meta-analysis. Rev Neurol (Paris) 176(7–8):560–570. https://doi.org/10.1016/j.neurol.2019.12.008
212. Hussein M, Fathy W, Abd Elkareem RM (2019) The potential role of serum vitamin D level in migraine headache: a case-control study. J Pain Res 12:2529–2536. https://doi.org/10.2147/JPR.S216314
213. Cayir A, Turan MI, Tan H (2014) Effect of vitamin D therapy in addition to amitriptyline on migraine attacks in pediatric patients. Braz J Med Biol Res 47(4):349–354. https://doi.org/10.1590/1414-431x20143606
214. Mottaghi T, Askari G, Khorvash F et al (2015) Effect of vitamin D supplementation on symptoms and C-reactive protein in migraine patients. J Res Med Sci 20(5):477–482. https://doi.org/10.4103/1735-1995.163971
215. Buettner C, Nir RR, Bertisch SM et al (2015) Simvastatin and vitamin D for migraine prevention: a randomized, controlled trial. Ann Neurol 78(6): 970–981. https://doi.org/10.1002/ana.24534
216. Gazerani P, Fuglsang R, Pedersen JG et al (2019) A randomized, double-blinded, placebo-controlled, parallel trial of vitamin D_3 supplementation in adult patients with migraine. Curr Med Res Opin 35(4): 715–723. https://doi.org/10.1080/03007995.2018.1519503
217. Taylor PN, Davies JS (2018) A review of the growing risk of vitamin D toxicity from inappropriate practice. Br J Clin Pharmacol 84(6):1121–1127. https://doi.org/10.1111/bcp.13573
218. Silva C, Fung AWS, Masson V et al (2023) Vitamin D toxicity from an unusual and unexpected source: a report of 2 cases. Horm Res Paediatr 96(3):332–340. https://doi.org/10.1159/000526755
219. Cui X, Eyles DW (2022) Vitamin D and the central nervous system: causative and preventative mechanisms in brain disorders. Nutrients 14(20): 4353. https://doi.org/10.3390/nu14204353
220. Asif A, Farooq N (2023) Vitamin D toxicity. In: StatPearls. StatPearls Publishing, Treasure Island
221. Gummin DD, Mowry JB, Beuhler MC et al (2022) 2021 annual report of the National Poison Data System© (NPDS) from America's poison centers: 39th annual report. Clin Toxicol (Phila) 60(12): 1381–1643. https://doi.org/10.1080/15563650.2022.2132768
222. Rastogi S, Agarwal A, Bhandari RL et al (2024) Hypervitaminosis D due to overdose: a case series. Indian J Pain 38:153–155. https://doi.org/10.4103/ijpn.ijpn_89_23
223. Ahuja A, Agrawal S, Daiya V et al (2024) Unraveling neurological shades: vitamin D Toxication and central pontine Myelinolysis exposed. Cureus 16(2): e53806. https://doi.org/10.7759/cureus.53806
224. Lima GO, Menezes da Silva AL, Azevedo JEC et al (2022) 100 years of vitamin D: supraphysiological doses of vitamin D changes brainwave activity patterns in rats. Endocr Connect 11(3):e210457. https://doi.org/10.1530/EC-21-0457
225. Kreienkamp R, Croke M, Neumann MA et al (2016) Vitamin D receptor signaling improves Hutchinson-Gilford progeria syndrome cellular phenotypes. Oncotarget 7(21):30018–30031. https://doi.org/10.18632/oncotarget.9065
226. Lam V, Takechi R, Mano J (2017) Vitamin D, cerebrocapillary integrity and cognition in murine model of accelerated ageing. Alzheimers Dement 13:1304. https://doi.org/10.1016/j.jalz.2017.06.1990
227. Dickens AP, Lang IA, Langa KM, Kos K, Llewellyn DJ (2011) Vitamin D, cognitive dysfunction and dementia in older adults. CNS Drugs 25(8):

629–639. https://doi.org/10.2165/11593080-000000000-00000
228. Demay MB, Pittas AG, Bikle DD et al (2024) Vitamin D for the prevention of disease: an endocrine society clinical practice guideline. J Clin Endocrinol Metab 109(8):1907–1947. https://doi.org/10.1210/clinem/dgae290
229. Vasil N, Razzaque MS (2025) Decoding vitamin D intake: how much is too much? Adv Exp Med Biol 1493:xx–xx. https://doi.org/10.1007/978-3-032-04357-3_2

Vitamin D and Mineral Ion Regulation

Hind S. Ahmed

Abstract

Mineral ion regulation and vitamin D activation are closely interconnected processes that are vital for maintaining overall health, particularly in terms of bone metabolism and calcium (Ca^{+2}) homeostasis. Mineral formation pathways involve the uptake of ions from the environment, the transport of ions by cells, sometimes temporary storage, and ultimately deposition inside or outside the cells. Minerals and trace elements are crucial micronutrients with recognized biochemical and biological roles. Insufficiency in these micronutrients is related to extensive human illnesses. This highlights the physiopathological aspects of the hormone-mediated control of biomineralization. Several component systems that affect bone formation, maintenance, and nutritional health, including the functions of Ca^{+2}, inorganic phosphate (Pi), dietary intake, and the renal excretion of minerals, are also highlighted. Additionally, there is an elusive balance between activators and inhibitors of mineralization, such as phosphorus (PO_4^{-3}), ionized Ca^{+2}, and pyrophosphate. Moreover, this chapter summarizes the significant importance of restricted regulation of serum Ca^{+2} and PO_4^{-3} levels in numerous biological processes, i.e., parathyroid hormone (PTH), vitamin D, vitamin K, and fibroblast growth factor 23 (FGF23). Finally, this chapter highlights the physiology and pathophysiological findings, clinical characteristics, and treatments.

Keywords

Calcium · Fibroblast growth factor 23 · Magnesium. Parathyroid hormone · Phosphate · Vitamin D · Zinc

Mineral Ion Regulation

Calcium and Phosphate Balance

Calcium and PO_4^{-3} are essential for bone health. The body regulates its levels through PTH and calcitonin. Calcium has many biological roles in the body, including neural, cardiac, and skeletal mineralization. It is the main bone constituent and provides the skeleton strength and structure. Both minerals are found in the bone, in serum, or intracellularly in soft tissues. Specifically, osteoblasts and osteoclasts use Ca^{+2} signals to regulate their differentiation and activity. Among the 1% of total body Ca^{+2} circulating in the blood, 40% is bound to protein, 10% is complexed with other anions, and the remaining 50% is free of ionized Ca^{+2}. The primary

H. S. Ahmed (✉)
Department of Chemistry, College of Education for Pure Science (Ibn Al-Haitham), University of Baghdad, Baghdad, Iraq
e-mail: hind.sh.a@ihcoedu.uobaghdad.edu.iq

M. S. Razzaque (ed.), *Vitamin D Function*, Advances in Experimental Medicine and Biology 1493,
https://doi.org/10.1007/978-3-032-04357-3_6

prominent metabolic function of the bone is to store minerals, particularly Ca^{+2} and PO_4^{-3}, with 99% of the body's Ca^{+2} found in the skeleton [1]. Enzyme activation, coagulation pathways, hormone and neurotransmitter release, and muscular contraction are all influenced by Ca^{+2} [2]. Serum soluble Ca^{+2} and PO_4^{-3} are important for maintaining homeostasis and the operation of the nervous, muscular, and other systems. As a result, hormonal controllers closely regulate intestinal absorption, bone resorption, renal excretion, and reabsorption. Nevertheless, bone integrity is not maintained by controlling Ca^{+2} homeostasis. Instead, the bone, parathyroid gland, intestine, and kidney form a multitissue axis to maintain serum Ca^{+2} within a narrow range (8.9–10.2) of mg/dL. Therefore, intestinal Ca^{+2} absorption following a meal is a signal that increases and disturbs serum Ca^{+2}, whereas bone formation and resorption, as well as renal Ca^{+2} excretion, react to serum Ca^{+2} influx in an effort to reduce serum Ca^{+2} disturbances [3]. Ca-sensing receptors (CaSRs), specifically G protein-coupled receptors in the kidneys and parathyroid glands, are crucial in regulating and tracking serum Ca^{+2} levels. When serum ionized Ca^{+2} is decreased, it is sensed by CaSR in the parathyroid glands, resulting in increased PTH secretion [4].

Inorganic phosphate is a vital mineral for life and plays crucial roles in various cellular processes, such as nucleic acid synthesis, energy storage, cell signal transmission, and bone formation. Phosphorus is an essential element that is important for numerous biological processes. Phosphorus-containing compounds are vital for maintaining cell structure and metabolism, generating ATP, regulating subcellular processes, maintaining acid-base homeostasis, and promoting bone mineralization. By maintaining a balance between osteoclast bone cells that create bone and osteoclasts that resorb bone, continuous mineralization and bone resorption preserve the equilibrium of PO_4^{-3} in the blood [5].

Serum Pi levels are closely controlled by various factors, including nutritional intake, intestinal absorption, the renal GFR, and tubular reabsorption. Pi is actively carried by type II sodium-dependent Pi cotransporters (NaPi-2b) in enterocyte brush border membranes and NaPi-2a and NaPi-2c in the brush border membrane (BBM) of renal proximal tubule cells [6]. Low dietary intake of Pi causes renal proximal tubule cells to express more PiT1. In the proximal tubules of the kidney, PiT2 is downregulated by the phosphaturic hormones PTH and FGF23 [7].

Healthy individuals typically have serum Pi values between 2.5 and 4.5 mg/dL. Serum Pi values greater than 4.5 mg/dL are considered hyperphosphatemia. There are three types of hypophosphatemia: mild (2–2.5) mg/dL, moderate (1–2) mg/dL, and severe (<1) mg/dL [8]. The global protein expression and protein phosphorylation patterns disrupted by extracellular Pi are yet unknown, although severe Pi-dysregulated AKT, FGF receptor (FGFR), and mitogen-activated protein kinase (MAPK) cell signaling pathways have been characterized. People often obtain Pi through their diet, which keeps their musculoskeletal systems functioning normally. Numerous illnesses, including cardiovascular disease (CVD), diabetes mellitus (DM), infertility, renal disease, and malignancies, can be caused by excessive consumption of dietary Pi, particularly from processed foods. High Pi causes pathological calcification, oxidative stress, cell death, and abnormal signal transmission, which are the molecular intermediates of these illnesses [9].

By binding to its ligand (RANKL), receptor activator of nuclear factor K (RANK) stimulates osteoclast precursors and increases osteoclast activity. When combined with macrophage colony-stimulating factor (M-CSF), osteocytes produce a ligand that activates osteoclasts and improves bone resorption. Osteoprotegerin, which is also produced by osteocytes, binds to RANKL and inhibits the activity of osteoclasts downstream [10]. Numerous cytokines and hormones influence this balance between osteoprotegerin secretion and RANK expression. This balance controls bone turnover and thus helps maintain the balance of intracellular PO_4^{-3} and Ca^{+2} [11]. Physiologically, these minerals are controlled by three dynamic hormones: vitamin D, PTH, and FGF23 [12].

Regulation by Vitamin D

Vitamin D plays a role in regulating the immune system. It controls suppressor T-cell activity, cytokine synthesis, and cell death [13]. In particular, vitamin D prevents the kidneys from excreting Pi and improves its absorption in the gut [14]. It is a lipid-soluble vitamin with a steroid structure, and its metabolite is known as 25-hydroxycholecalciferol [25(OH)D], which is converted to its active form, 1,25-dihydroxycholecalciferol [1,25(OH)$_2$D]. 7-Dehydrocholesterol may be converted to 25 (OH)D in the liver to produce vitamin D_3 from the sun [15], and 1,25(OH)$_2$D binds to the vitamin D receptor, forming heterodimers that initiate or inhibit gene transcription. When they enter the nucleus, they bind to vitamin D consent sequences, modifying gene transcription. Vitamin D (2 and 3) accumulates in adipose tissue [16, 17]. Consequently, metabolism to 25(OH)D by hydroxylases (cytochrome P450 (CYP) 2R1 and CYP27A1) occurs in the liver and other tissues in an autocrine and paracrine manner [18]. Hydroxylation occurs in renal tubules to form the 1,25(OH)$_2$D active molecule. Furthermore, two hydroxylases from the cytochrome P450 family, 1-α-hydroxylase (CYP27B1) and 24-α-hydroxylase (CYP24A1), are found in renal tubule cells. They convert 25(OH)D into either its inactive metabolite 24,25(OH)$_2$D or the active form of vitamin D (1,25(OH)$_2$D-calcitriol). Insulin-like growth factor 1, PTH, and low extracellular Ca^{+2} or Pi levels or ingestion promote the last phase. However, a variety of circumstances affect the efficacy of this combination [19], as shown in Fig. 1.

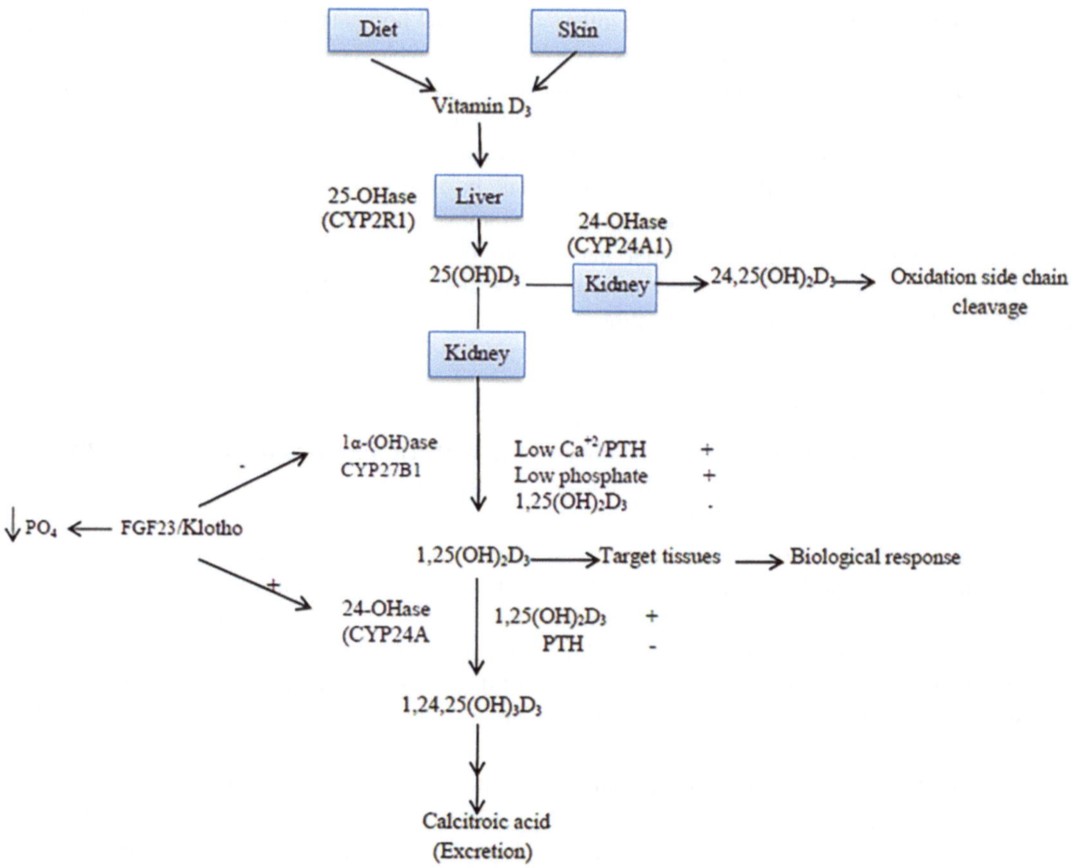

Fig. 1 Pathway of vitamin D metabolism

Genetic Factors Influencing Vitamin D Hemostasis

An association between vitamin D levels and epigenetic modifications has been recognized. A person's vitamin D level is influenced by their lifestyle and surroundings. However, genetic factors can also influence a person's blood vitamin D levels and, consequently, be linked to the genesis of certain chronic diseases, such as osteoporosis, cancer, and autoimmune disorders [20].

Genetic mutations and polymorphisms affect genes encoding vitamin D products, activators, transport proteins, VDRs, coactivating proteins, and protein modifications that interfere with proteins regulating vitamin D expression. The most frequently analyzed genes include DHCR7, CYP2R1, CYP27B1, GC, VDR, CYP24A1, and RXR [21]. Consequently, a defective conversion of calcidiol to calcitriol may be the consequence of inactivating mutations in the CYP27B1 gene, which are more often referred to as vitamin D-dependent rickets type 1A (VDDR1A) [22]. Less frequently, certain variations that impact CYP2R1 (VDDR1B) cause a defective 25-hydroxylation pathway, which prevents cholecalciferol from being converted to calcidiol. VDDR1B, or vitamin D-dependent rickets type 1B, is another name for this variation. These conditions result in hypocalcemia and aberrant bone mineralization by impairing the intestinal absorption of Ca^{+2} and Pi [23].

The mitochondrial enzyme CYP24A1 is found on chromosome 20 and is expressed in a number of target cells that have VDRs. As previously indicated, it catalyzes the degradation of 25 (OH)D and 1,25(OH)D. When vitamin D levels are low, this enzyme prolongs the half-life of 25 (OH)D and stops the buildup of harmful compounds [24]. Polymorphisms, including intronic SNVs such as rs222020, rs2282679, rs1155563, rs2298849, and rs222035, have been linked to decreased vitamin D expression and increased VDPB affinity. Other polymorphisms on the 30 edge area of the GC gene are also associated with vitamin D status [25]. The VDR gene, located on chromosome 12, encodes the vitamin D receptor. It has eight exons and six promoter regions. The ligand-binding domain binds 1,25(OH)D, while the DNA-binding domain interacts with target genes. Several hundred polymorphisms have been detected in the VDR gene, with TaqI, BsmI, ApaI, and FokI restriction endonucleases identifying the best-characterized sites. These SNPs are linked to various illnesses and homeostatic functions [26].

Low vitamin 25(OH)D levels are linked to plasma renin activity and angiotensin II concentrations, and supplemental cholecalciferol therapy may decrease renin-angiotensin-aldosterone system (RAAS) activity due to vitamin D deficiency [27]. The FokI polymorphism of the vitamin D receptor has been explored as a potential condition associated with an increased risk of arterial hypertension (HTN). It is thought to be connected to elevated renin and angiotensin II synthesis, hence accelerating the development of HTN [28]. Other complex genes are also considered to be indirectly involved in the control of vitamin D homeostasis and expression, although the signal is less clear. For example, CaSR receptor intracellular domain polymorphisms and cubilin polymorphisms have been studied; nevertheless, no correlations with vitamin D homeostasis have been reported [29]. One study that examined the impact of FGF23 gene variation on bone health and phosphate homeostasis revealed nine FGF23 polymorphisms, three of which are very prevalent: rs3832879, rs7955866, and rs11063112 [30].

Epigenetic Factors Influencing Vitamin D Hemostasis

Numerous studies have proposed a link between complex gene mutations in vitamin D metabolism and vitamin D status or a function for these mutations in vitamin D metabolism and a variety of disorders. Histone protein acetylation, methylation, and phosphorylation are the most prevalent genetic processes [31]. For example, it has been demonstrated that the methylation state of CYP2R1 controls how Ca^{+2} and vitamin D ingestion or radiation affect serum vitamin D levels. Additionally, it has been suggested that blood vitamin B12 and folate may play roles in controlling CYP27B1 and VDR methylation, respectively. The various effects of vitamin D supplementation on vitamin D biochemistry markers, genetic differences, and the transcriptional response of vitamin D target genes may be used to categorize a number of distinct groupings. Additionally, different molecular reactions in blood 25(OH)D levels have been reported for groups of poor, middle, and high responders to vitamin D administration [32].

Genomic Action

One nuclear receptor for steroid hormones that controls gene expression by acting as a ligand-activated transcription factor is the vitamin D receptor. The genomic mechanism of action of vitamin D is essential. 1,25(OH)D is among the best regulators of its activity. Unlike other nuclear receptors, the inner surface of VDR has a ligand-binding pocket that can surround and bind 1,25 (OH)D with an extremely high affinity of 0.1 nM [33]. In many human tissues and cell types, 1,25 (OH)D directly impacts the epigenome of more than 1000 genes by activating the vitamin D receptor, which causes changes in transcription and the proteome. Interestingly, in the cell nucleus, the vitamin D receptor exclusively targets the receptor. The regulatory mechanism of the target genes of vitamin D is currently not well described by a generic model. Once the vitamin D receptor has bound, it combines with several other nuclear receptors to form a multiprotein complex that preferentially binds to the DR3-type binding sites of promoter regions. The vitamin D receptors are common to this multiprotein complex.

Accordingly, the chromatin state is one of the many ways in which the vitamin D receptor epigenome controls the vitamin D receptor in target organs. It can work by regulating genes that encode chromatin up- or downregulated genes and by directly interacting with chromatin-modifying enzymes [34]. The architecture of many DNA genomic loops depends on vitamin D, and 1,25(OH)D affects the binding of chromatin regulatory proteins to several genes. As a result, 1,25(OH)D may alter the three-dimensional structure of chromatin. Each of these mediating processes may either increase or decrease RNA polymerase II activity and the mRNA expression of 1,25(OH)D target genes. Consequently, there is either an increase or a reduction in the expression of vitamin D target genes [35]. The expression of genes involved in both innate and adaptive immune activities can be modulated by vitamin D. Therefore, 25(OH)D and 1,25(OH)D have various effects on various immune cells, including B and T lymphocytes, monocytes, and macrophages [36]. Certain immunological inputs cause macrophages to express CYP27B1, which in turn causes 1,25 (OH)D to be locally produced at infection sites. This, in turn, directly increases the expression of genes that encode antimicrobial peptides. Vitamin D is involved in the modulation of a number of inflammatory or pro-inflammatory cytokines. Owing to its antiviral and antibacterial properties, 1,25(OH)D undoubtedly stimulates the production of cathelicidin through secretory processes. Additionally, as a paracrine agent, 1,25(OH)D stimulates nearby macrophages. The immunological tolerance of APCs can therefore be preserved, and they can control the surface expression of major histocompatibility complex (MHC) class II molecules, immunomodulatory cytokines, and costimulatory molecules [37]. Interleukin 10 (IL-10), which has anti-inflammatory effects, is expressed more frequently; however, IL-6 and IL-17, which have pro-inflammatory and atherogenic characteristics, are expressed less frequently. Furthermore, 1,25(OH)D interacts with metabolism and immune function by activating and modifying natural killer cells [38].

Emphasis on Clinical Outcomes

Research has shown that vitamin D has anticancer properties. This occurs directly by regulating the differentiation, proliferation, and death of neoplastic cells and indirectly by directing immune cells that are a part of the tumor microenvironment [39]. The risk of prostate, breast, and colorectal cancers is inversely correlated with circulating 25(OH)D levels, according to observational data, but not with the risk of cancer in general [40]. The actual impact of vitamin D on lowering cancer risk and enhancing patient health is still unclear [41, 42]. Multiple sclerosis, inflammatory bowel disease, and type 1 DM have all been linked to vitamin D insufficiency [43]. Therefore, there is still no evidence that vitamin D treatment may lower the incidence of these illnesses. In the context of the COVID-19 pandemic, vitamin D has recently been suggested to play a part. There are also contradictory findings from observational and interventional research. The risk and severity of COVID-19, together with higher rates of death and morbidity, are linked to low levels of circulating blood vitamin D. However, there is evidence of several confounding factors, which also depend on the specific circumstances of each patient, making the results untrustworthy. Furthermore, there is currently a dearth of trustworthy interventional evidence concerning vitamin D supplementation in hospitalized COVID-19 patients [44].

Regulation by PTH

The four parathyroid glands create PTH, which is essential for preserving the equilibrium of Ca^{+2} and PO_4^{-3} in the kidney, gut, and bone (Fig. 2). PTH increases the reabsorption of Ca^{+2} and PO_4^{-3} in the renal tubules, with increased excretion of PO_4^{-3}. In the intestine, PTH indirectly promotes the reabsorption of Ca^{+2} and PO_4^{-3} by increasing 1-α-hydroxylase [45].

PTH increases bone turnover and, eventually, bone resorption by stimulating osteoclasts [46]. The total result is an increase in serum Ca^{+2} and PO_4^{-3} and a decrease in serum PO_4^{-3} because bone reabsorption increases both Ca^{+2} and PO_4^{-3} levels. However, PTH also increases PO_4^{-3} renal excretion. Whereas PTH takes approximately 1–2 hours to start working in the bone, its effects on the kidney occur more quickly, within minutes [47].

Regulation by FGF23

FGF23 is an important hormone that controls PO_4^{-3}. It is believed to bind to the coreceptor Klotho and the receptor FGFR1 on the proximal renal tubule. When combined with this receptor, FGF23 inhibits 1-α-hydroxylase and lowers calcitriol levels. Elevated levels of Pi in the serum promote the synthesis of structural FGF23, which blocks Pi reabsorption through the renal tubules and calcitriol synthesis. Additionally, 24-α-hydroxylase, which converts 25 (OH)D into $24,25(OH)_2D$, an inactive metabolite, is increased by FGF23 [48]. The condition known as phosphatase rickets or hypophosphatasia causes renal Pi loss as a result of malfunctioning metabolic pathways that increase blood FGF23 levels.

1. The Na^+-dependent Pi transporter in the proximal renal tubule causes increased urine Pi loss when FGF23-activating mutations occur. The electrolyte imbalances and specific gene alterations are displayed in Table 1 [49].
2. PHEX (Pi-regulating endopeptidase homolog) on chromosome Xp22.2-p22.1 has a loss-of-function mutation that causes X-linked dominant hypophosphatemic rickets and is connected to elevated FGF23 production. Mature osteocytes and dentin cells often express PHEX, which contributes to a reduction in FGF23 expression [50].

Calcium and Phosphorus Disorders

Mutations in the vitamin D pathway may result in hyper- and hypocalcemia [51].

Hypercalcemia

Regular biochemical testing is typically used to identify hypercalcemia, which can manifest either abruptly or subtly. Increased thirst and polyuria, stomach and bone discomfort, volume loss, acute renal damage, depression, and kidney stones are

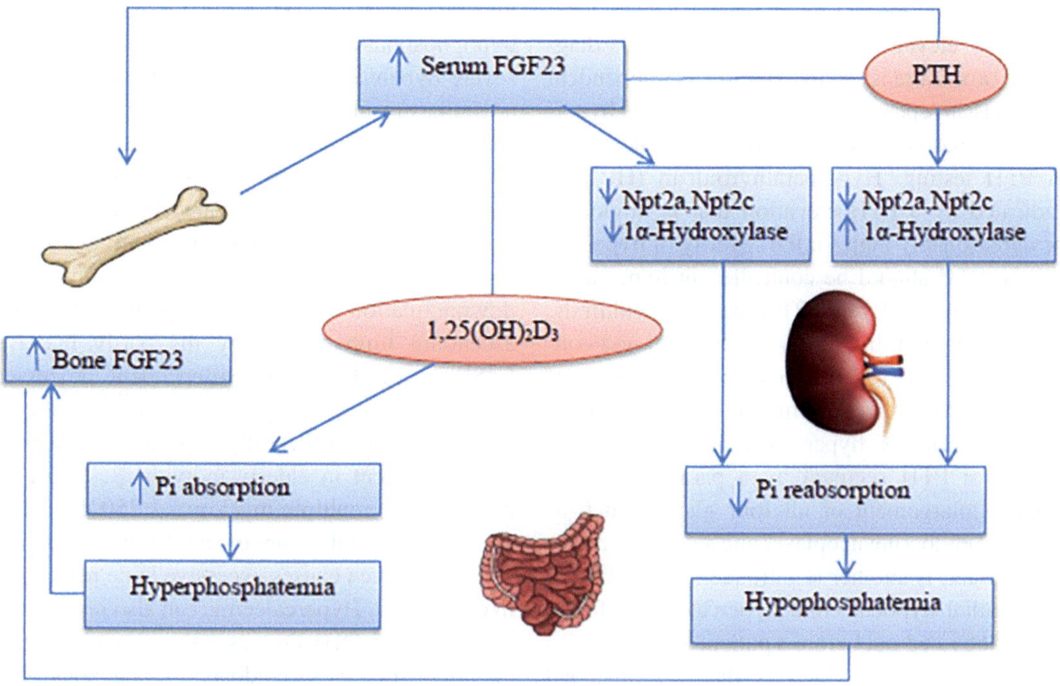

Fig. 2 Regulation of Ca-Pi homeostasis

Table 1 Signs and symptoms of hypercalcemia

Gastrointestinal (Abdominal moans)	Constipation and anorexia Abdominal pain Nausea and vomiting Pancreatitis Peptic ulcer disease
Renal (stones)	AKI CKD Nephrolithiasis Nephrogenic diabetes insipidus with resultant polyuria and polydipsia Renal tubular acidosis
Skeletal (bones)	Bone pain Fracture
Cardiovascular	Shortened QT interval Ventricular arrhythmias
Neurologic/psychiatric (psychic overtones)	Confusion Decrease LOC/altered MS Mood changes Coma

AKI acute kidney damage, *CKD* chronic kidney disease

some of the symptoms that patients may experience. Since hypercalcemia has several causes, a methodical approach to diagnosis is crucial [52] (Table 1).

The measurement of ionized Ca^{+2} is the first step in determining whether hypercalcemia is real; serum ALB control is not advised and is not reliable. While pseudohypercalcemia is less common than pseudohyponatremia is, it has been observed in multiple myeloma patients when elevated nonalbumin protein levels are linked to extracorporeal Ca^{+2}. On the other hand, the technique for measuring ionized Ca^{+2} is unaffected by elevated protein levels [53]. The diagnostic procedure is separated into PTH-dependent and PTH-independent processes when real hypercalcemia has been detected. The next crucial step is PTH testing. Hyperparathyroidism (HPT) is indicated by a PTH elevation that is obviously or unreasonably high for the severity of hypercalcemia; PTH should be controlled in hypercalcemic individuals. Even a PTH that falls within the normal range in this context is aberrant and consistent with HPT. The most frequent cause of PTH-induced hypercalcemia in the general population is primary hypercalcemia caused by an excess of PTH secreted by a parathyroid adenoma. Enlargement of all four glands, such as multiple endocrine neoplasia and seldom parathyroid cancer, is another less frequent cause [54].

Familial hypocalcemic hypercalcemia (FHH) should be ruled out before a patient is subjected to surgical resection, which is the recommended therapy for primary hypercalcemia caused by parathyroid adenoma. Neonatal severe HPT (NSHPT) is caused by loss-of-function mutations in both alleles, whereas FHH is caused by heterozygosity for loss-of-function mutations. Variants in CaSR have been proposed as potential explanations for human variations in bone mineral density and serum Ca^{+2} levels. Numerous studies have suggested that CaSR plays a role in controlling the administration of Ca^{+2} to the kidneys. Regardless of any apparent changes in the hormones that regulate Ca^{+2}, a higher level of extracellular Ca^{+2} results in a notable increase in the excretion of Ca^{+2} in the urine [55]. There is strong evidence that renal tubular CaSR regulates the reabsorption of divalent cations in both healthy and diseased states. Changes in serum Ca^{+2} may affect the excretion of Ca^{+2}, Mg^{+2}, and Na^{+} through a CaSR-dependent mechanism, according to clinical investigations conducted in people under PTH clamp circumstances. Because of these alterations, the Ca^{+2} set point increases, which keeps PTH secreted even in the face of hypercalcemia. To differentiate FHH from primary HPT, which results in hypercalciuria, increased PTH promotes tubular Ca^{+2} reabsorption, which results in hypocalciuria and moderate hypercalcemia. Hypermagnesemia and hypophosphatemia may also manifest. The more severe hypercalcemia in primary HPT caused by adenomas results in considerable increases in the clearance burden of Ca^{+2}, such that even with greater tubular reabsorption, hypercalciuria develops, even though elevated PTH stimulates renal Ca^{+2} reabsorption in both illnesses. Hypocalciuria also results from the downregulation of Ca^{+2} reabsorption in the thick ascending limb caused by the partly inactive CaSR in FHH. Non-PTH-dependent processes are the source of hypercalcemia in patients whose PTH is sufficiently repressed. Cancer is more prevalent in elderly individuals. It can be promoted by multiple myeloma, 1,25$(OH)_2$D, or the ectopic production of PTH-related peptide (PTHrP), which causes excessive bone resorption in osteoclasts. Hypercalcemia can also result from excess 1,25$(OH)_2$D, as in granulomatous disorders such as sarcoidosis or exogenous vitamin D poisoning [55].

Idiopathic Infantile Hypercalcemia An uncommon disorder known as idiopathic hypercalcemia in children (IIH) can have serious health repercussions if it is not identified in a timely manner [56]. Until 2011, the pathogenesis of IIH was unclear. Schlingmann discovered a mutation that caused the vitamin D-metabolizing enzyme 24 hydroxylase (CYP24A1) to lose its activity [57]. The liver is where 25-hydroxylase (CYP2R1) performs the first hydroxylation in the vitamin D activation cascade, resulting in 25 $(OH)_2$D [58]. In the kidney, the active form of 1,25-$(OH)_2$D is produced by 1-α-hydroxylase (CYP27B1), which completes the second hydroxylation. The five-step oxidation process of 1,25-$(OH)_2$D to soluble calcitroic acid is carried out by CYP24A1 through the inactivation of 24-hydroxylase (CYP24A1) [59]. Increased activity can result from loss-of-function mutations in CYP24A1 because of the buildup of active 1,25$(OH)_2$D. Numerous symptoms are present in affected individuals, including renal calcification, hypercalcemia, hypercalciuria, suppressed normal PTH levels, and elevated or abnormally normal serum 1,25$(OH)_2$D levels. A recessive mutation in the SLC34A1 gene, which codes for the proximal tubular NaPi-II, was later found, confirming the genetic diversity of IIH. These individuals exhibited hypophosphatemia as a result of renal potassium ion (K^+) squandering and average vitamin D in addition to hypercalcemia and PTH suppression [60].

The two types are distinguished by molecular heterogeneity: IIH type 1, which is due to CYP24A1 mutations, and IIH type 2, which is associated with SLC34A1 gene variations [61]. While heterozygotes might have an aberrant phenotype with varying symptoms at any age, homozygotes exhibit the clinical phenotype of IIH. Infants with IIH frequently experience lethargy, vomiting, polyuria, polydipsia, dehydration, constipation, poor feeding, weight loss, reduced muscle tone, and failure to flourish throughout the first year of life. Since heterozygous individuals exhibit a wide spectrum of symptoms, from moderate hypercalcemia to severe clinical and biochemical hypercalcemia, the type of inheritance of the condition is still under debate [62].

Lower FGF23 levels cause hypophosphatemia, which in turn results in decreased inhibition of 1-α-hydroxylase and decreased activity of 24-hydroxylase [63]. Increased calcitriol levels as a result of these mechanisms lead to hypercalcemia. A mutation that results in loss of function makes the 24-hydroxylase enzyme less active. The 24-hydroxylase enzyme breaks down 25(OH)D and $1,25(OH)_2D$ to create two inactive metabolites: $24,25(OH)_2D$ and $1,24,25(OH)_3D$. $1,25(OH)_2D$ builds up as a result of impaired degradation, which ultimately results in hypercalcemia and hypercalciuria [64].

Hypocalcemia

The symptoms of hypocalcemia range from mild and asymptomatic to severe and potentially fatal. Heart failure (HF), tetany, seizures, and QT prolongation can all result from acute hypocalcemia. Osteoporosis, dental cavities, calcification of the basal ganglia, cataracts, and other ocular symptoms can result from long-term hypocalcemia [65]. The physiological causes of hypocalcemia include Ca^{+2} sequestration, hypoparathyroidism, pseudohypoparathyroidism (PHP), and inadequate or resistant $1,25(OH)_2D$.

Another disorder that resembles hypoparathyroidism is severe hypomagnesemia, which is a necessary cofactor for PTH production and activity. Like hypercalcemia, ionized Ca^{+2} tests should be used to rule out PHP before depending on changes in the serum ALB concentration. Compared with hyponatremia or pseudohyperkalemia, pseudohypocalcemia is far less prevalent. Following the confirmation of genuine hypocalcemia, serum Mg^{+2} and PTH should be tested further; a Mg^{+2} shortage is proven to be the cause when hypocalcemia is corrected in response to Mg^{+2} delivery. Low or unsuitably "normal" PTH combined with hypocalcemia indicates HPT, whereas excessive PTH indicates PHP, vitamin D insufficiency, or CKD [66].

Hyperphosphatemia

Acute Hyperphosphatemia Patients with some degree of CKD who receive Pi-based enemas as bowel preparations for colonoscopy and develop acute nephropathy have been shown to experience severe AKI because of Ca-Pi deposition in the renal tubules and interstitium, which exacerbates hyperphosphatemia. Pi loading or the quick transfer of Pi from the inside to the extracellular space might result in acute hyperphosphatemia. Both pathways are worsened by a prior or concurrent decline in the GFR, which lowers filtration and, consequently, Pi excretion, which is further compromised by renal failure. Therefore, chronic hyperphosphatemia and acute alterations in Pi due to other causes are more common in individuals with severe CKD [67].

Chronic Hyperphosphatemia Because hypoparathyroidism or PHP and FGF23 deficiency or reduced effects of FGF23 enhance the expression of Na^+-Pi transporters in the PCT, which increases tubular reabsorption of filtered Pi, hyperphosphatemia results in the absence of CKD. Hypocalcemia is the main effect of PTH deficiency and occurs in conjunction with hyperphosphatemia when it is caused by deficiency. The family of drugs known as FGF receptor inhibitors (erdafitinib, pemigatinib, and infigratinib) that prevent the tubular effects of FGF23 include hyperphosphatemia.

Owing to the inability of FGF23 to reduce 1,25 $(OH)_2D$ levels, these medications and other disorders of FGF23 deficiency produce hypercalciuria and occasionally hypercalcemia, similar to disorders of elevated $1,25(OH)_2D$, such as granulomatous illnesses [68].

The injection of bisphosphonates and laxatives containing Na-Pi are other causes of drug-induced hyperphosphatemia. Because these medicines have the potential to cause AKI and life-threatening hyperphosphatemia, they should be administered extremely carefully in patients with CKD. Renal failure, the most frequent cause of chronic hyperphosphatemia, occurs when PTH-mediated bone resorption and ongoing dietary Pi absorption cause Pi to enter the bloodstream rather than be eliminated by the kidneys and filtered by conventional hemodialysis systems (daily intensive hemodialysis can remove significant amounts of Pi). The mechanism of hyperphosphatemia is determined by the patient's history, physical examination, and basic laboratory tests, such as the blood Ca^{+2} level; in certain cases, further testing for PTH, 1,25 $(OH)_2D$, and FGF23 levels is required for a conclusive diagnosis [69].

Although severe hypophosphatemia can result in cellular ATP depletion and musculoskeletal symptoms such as generalized skeletal muscular weakness, respiratory muscle weakness that can cause respiratory failure, and cardiac muscle weakness that can induce heart failure, mild hypophosphatemia is frequently asymptomatic. Proximal myopathy, dysphagia, intestinal blockage, rhabdomyolysis, hemolysis, and neurologic symptoms such as numbness, psychosis, generalized seizures, and coma are additional serious side effects. Depending on the age of initiation in relation to growth plate fusion, chronic hypophosphatemia can cause significant demineralization of bones and teeth, which can result in either osteomalacia or rickets [70].

Hypophosphatemia can be caused by redistribution of extracellular Pi into cells or loss of renal excretion. It is usually not caused by hypophosphatemia alone and can be confirmed by measuring 24-hour urinary Pi excretion or fractional Pi excretion in a spot urine test. Urinary Pi excretion >100 mg/day in hypophosphatemia is inappropriate and confirms renal Pi wasting. Fractional Pi excretion should be low in hypophosphatemia, and an increase of $>20\%$ confirms renal Pi wasting. Most chronic hypophosphatemia cases are due to renal Pi wasting, either due to increased FGF23 or PTH. Hypophosphatemia due to renal PTH wasting in the absence of hypercalcemia indicates primary FGF23 hyperfunction, which has several genetic and acquired causes [71]. Patients with malignancies and those receiving extensive hemodialysis may also experience hypophosphatemia as a result of increased Pi clearance. In these situations, Pi supplementation efficiently cures hypophosphatemia.

Hyperparathyroidism

Primary HPT is a disorder of the parathyroid glands, also called the parathyroid glands. In primary HPT, the parathyroid glands are hyperactive in one or more cases. The gland thus secretes an excessive amount of PTH, ultimately resulting in hypophosphatemia and hypercalcemia. This may occur alone or as a component of a hereditary condition [72]. PTH levels are evaluated by the level of CaSR, which senses extracellular serum Ca^{2+} levels. CaSR signals increase PTH synthesis and release when the extracellular serum Ca^{+2} concentration is at the lower end of the typical range. CaSR decreases PTH secretion when the extracellular serum Ca^{+2} concentration is at the higher end of the typical range. The "set point" is the moment at which CaSR sensing of the serum

Ca^{+2} concentration results in half-maximal PTH production.

An important aspect of the pathophysiology of HPT or hypoparathyroidism is the capacity of CaSR to promote PTH production on the basis of the set point [73]. Hypercalcemia is caused by mutations in CaSR that cause loss of function, which increases the serum Ca^{+2} concentration. Increased renal reabsorption of blood Ca^{+2} from tubular fluid and hypocalciuria are the results of the inability of Ca^{+2} to activate the receptor in the renal tubules [74].

Neonatal Severe Hyperparathyroidism and Familial Hypocalciuric Hypercalcemia

1. Owing to loss-of-function mutations in CaSR, type 1 FHH manifests as moderate hypercalcemia, elevated or aberrant PTH levels, and a 24-hour urine Ca^{+2} concentration <6.25 mmol or a urinary Ca^{+2}/creatinine clearance ratio <0.01 [75].
2. Mutations that cause GNA11 (guanine nucleotide-binding protein, alpha-11) to lose its function are the cause of type 2 FHH. Hypercalcemia is caused by the breakdown of a G protein signal transduction pathway that it encodes, which connects CaSR to intracellular signal transduction pathways [76].
3. Mutations that cause AP2S1 (adapter-associated protein complex 2, sigma1 subunit) to lose its function are the cause of type 3 FHH. Furthermore, inactivating mutations lead to poor autophagy of the CaSR receptor from the cell surface and faulty connections between adaptor-associated protein complex 2 and β-arrestin (ARRB1). Cognitive impairment, reduced bone mineralization, and hypercalcemia are among the clinical manifestations [77].
4. An inactivating mutation in the Ca2+ receptor causes neonatal severe HPT (NSHPT). However, it is more severe than HPT. When two aberrant Ca^{+2} receptor alleles are inherited in either the homozygous or compound heterozygous form, NSHPT results, which causes severe hypercalcemia at birth [78].

Isolation of Nonsyndromic Mutations in the CDC73, CaSR, GNA11, and AP2S1 Genes from Hyperparathyroidism Patients

1. The benign family variant is called type 1.
2. An increased incidence of parathyroid cancer, renal tumors, uterine tumors, and severe malignant fibro-osseous tumors of the maxilla and/or mandible is a symptom of type 2 tumors, also known as HPT-jaw tumor syndrome (HPT-JT) [79]. Table 2 shows the genetic variants that cause HPT.

Multiple Endocrine Neoplasia (MEN) Syndromes

1. Parathyroid hyperplasia, pituitary tumors (prolactinoma), and pancreatic tumors (gastrinoma, insulinoma, etc.) are among the cancers that constitute MEN1. The cause of MEN1 is an inactivating mutation in the Menin tumor suppressor gene, which is located on chromosome 11q13.1. MEN1 syndrome usually manifests in the second decade of life and is seldom observed before the age of 10. The first symptom is HPT [80].
2. Germline mutations in the RET gene, which codes for a transmembrane tyrosine kinase receptor essential for cell proliferation, differentiation, and survival, cause MEN2A. Pheochromocytoma, medullary thyroid cancer, and parathyroid adenomas are associated with this disease. Additionally, medullary thyroid cancer produces greater amounts of calcitonin, pheochromocytoma produces higher levels of serum or urine catecholamines, and hypercalcemia is a characteristic [81].
3. MEN4 and MEN1 share clinical similarities. On chromosome 12p13.1, it is generated by a deactivating mutation in cyclin-dependent kinase inhibitor 1B (CDKC1B). The tumor

Table 2 Genetic mutations causing HPT

Disease type of inheritance	Gene defect/ location	Serum PTH	Serum Ca^{+2}	Serum Pi	Serum 1,25 $(OH)_2D$	Urine Ca/Cr
Familial isolated HPT AD	*MEN1/menin* 11q13.1 1q31.2 *CaSR* 3q13.3-q21.1 *GCM2* 6p24	↑	↑	↓	↑	↑
HPT-jaw tumor syndrome AD	*CDC73* 1q31.2	↑	↑	↓	↑	↓
FHH type 1 AD	*CaSR* 3q13-q21	↑	↑	↓	↑	↓
FHH type 2 AD	*GNA11* 19p13.3	↑	↑	↓	↑	↓
FHH type 3 AD	*AP2S1* 19q13.32	↑	↑	↓	↑	↓
Neonatal severe HPT AR	*CaSR* Homozygous Inactivating	↑	↑↑↑	↓	↑	↓
MEN1 AD	*MEN1* or *menin* 11q13.1	↑	↑	↓	↑	↓
MEN2A AD	*RET* 10q11.2	↑	↑	↓	↑	↓
MEN4 AD	*CDKN1B*	↑	↑	↓	↑	↓

FGF23 fibroblast growth factor23, *1,25(OH)₂D* 1,25-dihydroxyvitamin D, *urine Ca/Cr* urine calcium/creatinine ratio, *Ca^{+2}* calcium, *Pi* phosphate

suppressor gene CDKN1B produces p27, which controls cell motility and death by stopping the cell cycle in the G0/G1 phase. Like MEN1, MEN4 is characterized by pituitary and parathyroid adenomas and may present with insulinomas, gastrinomas, or genitourinary tract cancers. Clinical manifestations of HPT are associated with unsuppressed PTH levels, hyperphosphatemia, and hypercalcemia. Metabolic results include increased normal serum PTH, decreased renal tubular reabsorption of Pi, increased calcitriol levels, hypercalcemia, hypophosphatemia, and hypermagnesemia. The presence of HPT syndrome necessitates genetic testing [82].

Hypoparathyroidism

A collection of conditions known as primary hypoparathyroidism are caused by a lack of PTH production, which results in hyperphosphatemia and hypocalcemia [83].

Isolated Hypoparathyroidism

1. Gain-of-function mutations in Ca^{+2} receptors cause type 1 hypocalcemia by increasing the sensitivity and responsiveness of the receptor to blood Ca^{+2} levels, which in turn results in abnormally low PTH production. This could be linked to type 5 Bartter syndrome, which is caused by the inability of the renal tubules to reabsorb sodium chloride.
2. Mutations in GNA11 (guanine nucleotide-binding protein alpha-11), which often initiates intracellular signal transduction from CaSR and Ca^{+2} binding, cause type 2 autosomal dominant hypocalcemia [84].

Familial Isolated Hypoparathyroidism

1. Both autosomal dominant and autosomal recessive PTH mutations in the 11p15.3 gene cause increased neuromuscular irritability as a result of Ca^{+2} deficiency, which manifests as seizures, tetany, and paresthesia [85].

2. GCM2 (glial cell missing 2) mutations on 6p24.2 encode a transcription factor necessary for parathyroid differentiation. Alkaline phosphatase activity is unaffected by hypoparathyroidism, despite low blood Ca^{+2} and 1,25 $(OH)_2D$ levels, elevated Pi levels, and low or undetectable PTH levels. Intestinal Ca^{+2} absorption and bone resorption are inhibited, despite an increase in fractional Ca^{+2} excretion. Tubular Pi reabsorption is increased, and renal cyclic AMP excretion, 24-hour urine Ca^{+2} excretion, and renal clearance of the Ca^{+2} load are reduced [85].

Parathyroid Hormone-Pseudohypoparathyroidism

End-organ resistance to PTH is a characteristic of a diverse set of illnesses known as PHP. The pathophysiology involves features of hyperphosphatemia and hypocalcemia, although it is different from that of hypoparathyroidism.

1. The most prevalent form of PHP, PHP1A, is caused by deactivating maternal GNAS mutations. The α-subunit of the catalytic guanine nucleotide-binding protein Gsα is encoded by the GNAS gene. Both maternal and paternal alleles are expressed in other organs, but only the maternal allele is expressed in the renal tubules [86].
2. Unlike PHP1A, PHP is caused by a mutation in the paternal allele of the GNAS gene rather than the maternal allele. These individuals only have AHO, but because the maternal allele is still expressed in the renal tubules, they do not have any biochemical modifications of the PHP1A gene, such as PTH resistance [86].
3. Isolated renal resistance to PTH, which arises from methylation abnormalities in differentially methylated regions (DMRs) of the maternal allele of GNAS, leads to reduced production of Gsα and PHP1B. Owing to their normal phenotype, absence of AHO results, and typically absence of other endocrine abnormalities, these patients are sometimes challenging to identify. Owing to their similar biochemistry, PHP1A and PHP1B cause hypophosphatemia, hyperphosphatemia, and HPT [87].
4. A variable GNAS mutation in exon 13 causes poor Gsα receptor interaction, which results in PHP1C. The biochemical and clinical characteristics of these abnormalities are comparable to those of PHPIA [88].
5. The catalytic component of adenylate cyclase is encoded by PRKAR1A (protein kinase, cAMP dependent, regulatory, type 1, alpha), and mutations in this gene cause PHP2 [89].

Magnesium

Many foods naturally contain Mg^{+2}, which may also be purchased as a dietary supplement and is contained in certain drugs, including laxatives and antacids. However, because of modern agriculture and changes in lifestyle, the amount of Mg^{+2} consumed from natural foods has decreased in recent decades. Although the necessary quantity increases during pregnancy, the recommended daily allowance (RDA) of Mg^{+2} for adults is between 310 and 420 mg/dL (Table 3) [90].

The body's coordinated interactions between nutrients are vital for several biochemical processes and affect how PTH and vitamin D function; Ca^{+2} metabolism may be hampered by low Mg^{+2} levels. $1,25(OH)_2D$ has been shown to promote the intestinal absorption of Mg^{+2} [91].

In type 2 diabetic individuals aged 53.6 ± 10.7 years, the effects of vitamin D supplementation on circulating levels of Mg^{+2} were examined; following 6 months of vitamin D_3 treatment (2000 IU/d), a substantial increase in blood levels of Mg^{+2} was observed [92]. Furthermore, the vitamin D-binding protein requires Mg^{+2} as a cofactor. Moreover, as previously stated, vitamin D metabolism into the active form of $1,25(OH)_2D$ is a Mg^{+2}-dependent process that involves renal 1α-hydroxylation and hepatic 25-hydroxylation. Its absence has been linked to Mg^{+2}-dependent vitamin D-resistant rickets and

Table 3 RDA of Mg^{+2} [90]

Age	Amount of Mg^{+2} (mg/dL)		
	Male	Female	Pregnancy
≤6 months	30	30	–
7–12 months	75	75	–
1–3 years	80	80	–
4–8 years	130	130	–
9–13 years	240	240	–
14–18 years	410	360	400
19–30 years	400	310	350
31–50 years	420	320	360
≥51 years	420	320	–

causes decreased $1,25(OH)_2D$ levels and a compromised PTH response. The resistance to vitamin D treatment was significantly decreased by magnesium supplementation [93].

As the second most prevalent intracellular cation, it influences the formation of active vitamin D metabolites, which are essential for bone mineralization. When the Mg^{+2} equilibrium is not maintained in the body, the efficacy and therapeutic advantages of vitamin D are greatly diminished. Additionally, vitamin D is essential for the intestinal absorption of Mg^{+2} and Pi, which ultimately affects the process of bone mineralization. The three main enzymes that convert vitamin D and the proteins that bind it rely on magnesium. These enzymes include the liver's 25-hydroxylase and the kidneys' 1α-hydroxylase and 24-hydroxylase. Therefore, in rickets patients, Mg^{+2} supplementation significantly restored resistance to vitamin D therapy [94]. Additionally, Mg^{+2} is important for immunoregulation. The effects of vitamin D metabolites play crucial roles in immunocompetence as well as natural and adaptive immunity [95]. Additionally, Mg^{+2} ingestion altered the possible relationships between serum 25(OH)D and mortality, especially in patients with colorectal cancer and CVD, and the inverse connections were mostly observed in those whose Mg^{+2} intake was higher than the median. Magnesium is essential for keeping the heart healthy because it helps keep the heart's beat steady and helps keep the heart from clotting abnormally. Additionally, it is very successful in lowering the incidence of heart attacks and strokes and aids in maintaining normal blood pressure levels [96].

Bone mineral density and dietary Mg^{+2} consumption have been found to be positively correlated. Nonetheless, the majority of research on osteoporosis prevention and therapy focuses on the consumption of more Ca^{+2} and vitamin D. The size of the hydroxyapatite crystals in the bone is determined by the binding of Mg^{+2} to their surface. More importantly, crystals in Mg^{+2}-deficient bones can result in brittle, fracture-prone bones; Mg^{+2} aids in osteoblast proliferation in addition to skeletal mineralization, and its absence hinders the production of new bone [97]. It causes vitamin D insufficiency, lowered PTH levels, and the development of osteoporosis in individuals. A decrease in Mg^{+2} levels may result in decreased PTH production and secretion as well as a smaller number of existing vitamin D receptors in target cells, since it has been proposed that Mg^{+2} may affect PTH synthesis and control the amount of VDR [98].

Zinc

The natural sources of zinc (Zn^{+2}) include meals, medications, and dietary supplements. It is a necessary trace mineral that is found in more than 600 enzymes, different proteins, and more than 2500 transcription factors [99]. Numerous biological functions, such as cellular death, the immunological response, brain growth and maturation, taste and smell modulation, skin and mucosal integrity, and metabolic function, depend on Zn^{+2}. It is a signaling ion that regulates gene expression, protein synthesis, DNA synthesis, and cell division. It is attached to plasma proteins such as metallothionein, transferrin, and albumin. Although the plasma Zn^{+2} level has been tested for assessing Zn^{+2} status, it is not a reliable indicator of the body's overall Zn^{+2} content. This is partly caused by the fluctuating Zn^{+2} levels of the intracellular compartments, which include nucleic acids and are unable to be assessed in plasma samples [100]. Through gastrointestinal excretion in the feces and temporary storage in intestinal cells before being sloughed off to be expelled, Zn^{+2} homeostasis is maintained. Numerous health risks, such as obesity, metabolic disorders, HTN, and other CVDs, are linked to zinc deficiency. There is additional evidence linking the development of COVID-19 to low Zn^{+2} levels [101].

Reduced absorption, increased nutritional demand, excessive excretion, or insufficient dietary intake can contribute to deficits in Zn^{+2} and vitamin D. Because of its many extraskeletal uses, vitamin D deficiency can be linked to a number of cancers, CVD, immunological disorders, metabolic syndrome, obesity, DM, and renal disorders [102].

Zinc and Vitamin D Interactions

For vitamin D to activate and perform its functions in immunity and bone health, a number of cofactors are needed. Among these is Zn^{+2}, which is a necessary component for the proper operation of vitamin D, and Zn^{+2} is required for $1{,}25(OH)_2D$ to have biological effects. Like other steroid receptors, the VDR has the ability to selectively engage with Zn^{+2} finger regions. These ions are present in numerous transcription factors, including those found in the VDR, and include one or two Zn^{+2} ions bound by cysteine or histidine residues [103]. For $1{,}25(OH)_2D$ to have biological effects, Zn^{+2} is necessary. The VDR can preferentially interact with Zn^{+2} finger regions, similar to other steroid receptors. These consist of one or two Zn^{+2} ions bound by cysteine or histidine residues and are present in several transcription factors, including those identified in the VDR. Consequently, Zn^{+2} is bound by the VDR, and the function of vitamin D-dependent genes is influenced by the state of Zn^{+2}. Eight weeks of Zn^{+2} supplementation increased postmenopausal women's vitamin D levels in a double-masked, randomized study [104]. Hereditary vitamin D-resistant rickets (HVDRRs), which rarely occur in children, are autosomal recessive conditions characterized by abnormalities in bone mineral structure. Defective intestinal absorption of Ca^{+2} resulting from VDR mutations in the gastrointestinal system is the primary effect of HVDRR.

A point mutation in the first Zn^{+2} finger of the VDR DNA-binding domain was discovered in a Brazilian investigation involving two affected brothers. The amino acid arginine changes to a stop codon because of a single base change caused by this mutation. A shortened protein and the deletion of a portion of the Zn^{+2} fingers and ligand-binding domain were the results of the nonsense mutation [105]. Similar symptoms caused by the same mutation were found in a different investigation of a young French-Canadian child [106]. The mutant allele was heterozygous in the parents of the patients. A diverse missense mutation occurred in the first Zn^{+2} finger of the DNA-binding domain in a Chinese study, with similar skeletal effects [107].

The importance of the Zn^{+2} finger-binding domain with the VDR and its influence on Ca^{+2} absorption and bone mineralization have been demonstrated by these investigations. In the absence of Zn^{+2} fingers, resistance to $1{,}25(OH)_2D$ binding prevents the proper transcription of vitamin D-dependent genes, preventing any downstream effects of this ligand and resulting in symptoms in HVDRR patients. When the VDR bound to Zn^{+2}, conformational changes were observed inside the DNA-binding domain. Through the induction of Zn^{+2} transporters, vitamin D can affect cellular Zn^{+2} homeostasis. This correlation indicates that Zn^{+2} homeostasis inside cells can be regulated by vitamin D [108].

Vitamin K

There are two varieties of vitamin K: menaquinones (vitamin K2) and phylloquinones (vitamin K1). They are cofactors for a carboxyl moiety that is added posttranslationally by γ-carboxylation to glutamate residues that are attached to proteins. Because it creates a complex Ca-binding site in the coagulation cascade, the γ-carboxylation of glutamate residues (carboxyglutamic acid) or Gla residues is essential for prothrombin activity. Vitamin K-dependent Gla residues are found in approximately 17 additional proteins, such as osteocalcin [109]. The most prevalent noncollagenous component in the mineralized bone matrix is osteocalcin. In its fully carboxylated form, osteocalcin prevents calcification. On the other hand, it seems that the undercarboxylated version of osteocalcin regulates both reproductive and energy metabolism [110].

Vascular smooth muscle cells produce the extrahepatic vitamin K-dependent matrix protein Gla, which efficiently prevents arterial and soft tissue calcification. In the development plate, overexpression of the matrix Gla protein prevents cartilage calcification and endochondral ossification. Inhibitors, such as matrix Gla protein, prevent Ca^{+2} and Pi from precipitating spontaneously because they are present in supersaturated amounts in extracellular fluids [111]. Furthermore, arterial calcification is linked to the inhibition of matrix Gla protein carboxylation by vitamin K antagonists. The bone and calcified cartilage include additional carboxylated protein, which is a special cartilage matrix-associated protein. Thus, an increased likelihood of soft tissue calcification may aggravate vitamin K shortages [112]. Vitamin K-dependent proteins are decarboxylated by coumarins, which are vitamin K antagonists that prevent vitamin K recycling and decrease tissue vitamin K reserves. Compared with vitamin K-independent anticoagulants, warfarin is linked to a greater rate of arteriosclerosis and coronary artery calcification [113].

Therapeutic Approaches

Although the foundation of treatment for hereditary abnormalities in Ca^{+2} metabolism differs depending on the specific disease, it is often the same. The intensity and clinical trajectory determine whether early therapy is necessary. Stabilizing serum Ca^{+2} levels within the normal range is an urgent therapeutic objective to prevent seizures and potentially lethal cardiac arrhythmias. The objective is to maintain proper hydration and increase Ca^{+2}, PO_4^{-3}, and Mg^{+2} levels after stabilization. Confirming optimal bone mineralization and preventing problems and long-term therapeutic adverse effects, such as kidney stones, are the following steps. Furthermore, if clinically necessary, determining the etiology and performing genetic analysis are crucial [114].

If a patient does not react to vitamin D treatment, it may be believed that vitamin D and vitamin D-dependent rickets (VDDRs) coexist. Conventional vitamin D supplementation dosages might prevent nutritional rickets [115], but VDDR requires far greater dosages. Oral sodium phosphate supplementation is another factor that might impact Ca^{+2}-PO_4^{-3} homeostasis, especially in patients with nephrolithiasis; following sodium phosphate therapy, these patients have more pronounced hypocalciuria and increased PTH levels [116].

Conclusions

Mineral ion regulation and proper vitamin D activation are essential for maintaining bone health and overall metabolic functions. Imbalances can lead to numerous illnesses, i.e., DM, CVD, impaired metabolism, and bone disorders. Frequently monitoring Ca^{+2}, Pi, PTH, FGF23, vitamin D, and Mg^{+2} levels is important for ideal health. Adipose tissue produces, stores, and breaks down the active form of vitamin D, according to data analysis. Additionally, adipocytes contain VDR, which enables vitamin D to influence adipose tissue in both genetic and nongenomic ways. Adipocyte secretion, oxidative stress, inflammation, apoptosis, lipid metabolism, and thermogenesis are all impacted by vitamin D. Consequently, vitamin D preserves the lipid content, shape, and function of adipose tissue. Consequently, it is hypothesized that vitamin D supplementation might be a viable treatment option for the malfunctioning adipose tissue associated with obesity and type 2 DM. To ascertain the advantages of vitamin D supplementation in people with different degrees of metabolic problems and obesity, more clinical research is necessary. Therefore, a lack of vitamin K increases the likelihood of soft tissue and artery calcification, whereas a lack of vitamin D is linked to poor bone mineralization and cartilage. Future research into the ways in which vitamin D and vitamin K affect soft tissue calcification might yield crucial information for developing novel strategies to stop atherosclerosis and calcification of the artery wall.

References

1. Vannucci L, Fossi C, Quattrini S et al (2018) Calcium intake in bone health: a focus on calcium- rich mineral waters. Nutrients 10(12):1930. https://doi.org/10.3390/nu10121930
2. Root A, Levine M (2021) 20—disorders of mineral metabolism ii. Abnormalities of mineral homeostasis in the newborn, infant, child, and adolescent. In: Sperling MA (ed) Sperling pediatric endocrinology, 5th ed. Elsevier, Philadelphia, pp 705–813. https://doi.org/10.1016/B978-0-323-62520-3.00020-8
3. Bronner F, Aubert J (1965) Bone metabolism and regulation of the blood calcium level in rats. Am J Phys 209(5):887–890. https://doi.org/10.1152/ajplegacy.1965.209.5.887
4. Root A (2018) Genetic disorders of calcium, phosphorus, and bone homeostasis. Transl Sci Rare Dis 3(1):1–36. https://doi.org/10.3233/TRD-180019
5. Kliegman R, Geme J, Blum N et al (2020) Electrolyte and acid-base disorders. In: Nelson textbook of pediatrics. Elsevier, Philadelphia, pp 389–425
6. Goodson JM, Shi P, Razzaque MS (2019) Dietary phosphorus enhances inflammatory response: a study of human gingivitis. J Steroid Biochem Mol Biol 188:166–171. https://doi.org/10.1016/j.jsbmb.2019.01.023
7. Tomoe Y, Segawa H, Shiozawa K et al (2010) Phosphaturic action of fibroblast growth factor 23 in Npt2 null mice. Am J Physiol Ren Physiol 298(6): F1341–F1350. https://doi.org/10.1152/ajprenal.00375.2009
8. Christopoulou EC, Filippatos TD, Megapanou E et al (2017) Phosphate imbalance in patients with heart failure. Heart Fail Rev 22(3):349–356. https://doi.org/10.1007/s10741-017-9615-6
9. He P, Mann-Collura O, Fling J et al (2021) High phosphate actively induces cytotoxicity by rewiring pro-survival and proapoptotic signaling networks in HEK293 and HeLa cells. FASEB J 35(1):e20997. https://doi.org/10.1096/fj.202000799RR
10. Uday S, Högler W (2019) Rickets and osteomalacia. In: Huhtaniemi I, Martini L (eds) Encyclopedia of endocrine diseases, 2nd ed. Academic Press, Oxford, UK, pp 339–354
11. Tahir NT, Ahmed HS, Ahmed HS (2024) Studies on the role of retinol binding protien-4 in type 2 diabetic Iraqi patients with metabolic syndrome. Nucleus (India) 67(2):351–357. https://doi.org/10.1007/s13237-023-00430-w
12. Razzaque MS (2011) Phosphate toxicity: new insights into an old problem. Clin Sci 120(3):91–97. https://doi.org/10.1042/CS20100377
13. Rebelos E, Tentolouris N, Jude E (2023) The role of vitamin D in health and disease: a narrative review on the mechanisms linking vitamin D with disease and the effects of supplementation. Drugs 83(8):665–685. https://doi.org/10.1007/s40265-023-01875-8
14. Akimbekov N, Digel I, Sherelkhan D et al (2022) Vitamin D and phosphate interactions in health and disease. Phosphate Metab Physiol Toxic 1362:37–46. https://doi.org/10.1007/978-3-030-91623-7_5
15. Tahir NT, Ahmed HS, Hashim RK et al (2020) The correlation between vitamin D and clinical implications for obesity-related type 2 diabetes mellitus. JOBRC 14(1): 39–45. https://doi.org/10.24126/jobrc.2020.14.1.587
16. McDonnell D, Mangelsdorf D, Pike J et al (1987) Molecular-cloning of complementary-DNA encoding the avian receptor for vitamin-D. Science 235(4793):1214–1217. https://doi.org/10.1126/science.3029866
17. Park C, Han S (2021) The role of vitamin D in adipose tissue biology: adipocyte differentiation, energy metabolism, and inflammation. J Lipid Atheroscler 10(2):130– 144. https://doi.org/10.12997/jla.2021.10.2.130
18. Altieri B, Cavalier E, Bhattoa H et al (2020) Vitamin D testing: advantages and limits of the current assays. Eur J Clin Nutr 74(2):231–427. https://doi.org/10.1038/s41430-019-0553-3
19. Erem A, Razzaque M (2021) Vitamin D-independent benefits of safe sunlight exposure. J Steroid Biochem Mol Biol 213:105957. https://doi.org/10.1016/j.jsbmb.2021.105957
20. Underland L, Markowitz M, Gensure R (2020) Calcium and phosphate hormones: vitamin D,

parathyroid hormone, and fibroblast growth factor 23. Pediatr Rev 41(1):3–11. https://doi.org/10.1542/pir.2018-0065
21. Bahrami A, Sadeghnia HR, Tabatabaeizadeh S et al (2018) Genetic and epigenetic factors influencing vitamin D status. J Cell Physiol 233(5):4033–4043. https://doi.org/10.1002/jcp.26216
22. Lips P (2007) Relative value of 25(OH)D and 1,25 $(OH)_2D$ measurements. J Bone Miner Res 22(11): 1668–1671. https://doi.org/10.1359/jbmr.070716
23. Demir K, Kattan WE, Zou M et al (2015) Novel CYP27B1 gene mutations in patients with vitamin D-dependent rickets Type 1A. PLoS One 10(7): e0131376. https://doi.org/10.1371/journal.pone.0131376
24. Datta P, Philipsen PA, Olsen P et al (2017) The half-life of 25(OH)D after UVB exposure depends on gender and vitamin D receptor polymorphism but mainly on the start level. Photochem Photobiol Sci 16(6):985–995. https://doi.org/10.1039/c6pp00258g
25. Nissen J, Rasmussen LB, Ravn-Haren G (2014) Common variants in CYP2R1 and GC genes predict vitamin D concentrations in healthy Danish children and adults. PLoS One 9(2):e89907. https://doi.org/10.1371/journal.pone.0089907
26. Usategui-Martín R, De Luis-Román D-A, Fernández-Gómez JM et al (2022) Vitamin D receptor (VDR) gene polymorphisms modify the response to vitamin D supplementation: a systematic review and meta-analysis. Nutrients 14(2):360. https://doi.org/10.3390/nu14020360
27. Carrara D, Bernini M, Bacca A et al (2014) Cholecalciferol administration blunts the systemic renin–angiotensin system in essential hypertensives with hypovitaminosis D. J Renin-Angiotensin-Aldosterone Syst 15(1):82–87. https://doi.org/10.1177/1470320312471149
28. Shree T, Banerjee P, Senapati S (2023) A meta-analysis suggests the association of reduced serum level of vitamin D and T-allele of Fok1 (rs2228570) polymorphism in the vitamin D receptor gene with celiac disease. Front Nutr 9:996450. https://doi.org/10.3389/fnut.2022.996450
29. Tourkochristou E, Mouzaki A, Triantos C (2023) Gene polymorphisms and biological effects of vitamin D receptor on nonalcoholic fatty liver disease development and progression. Int J Mol Sci 24(9): 8288. https://doi.org/10.3390/ijms24098288
30. Pekkinen M, Laine CM, Mäkitie R et al (2015) FGF23 gene variation and its association with phosphate homeostasis and bone mineral density in Finnish children and adolescents. Bone 71:124–130. https://doi.org/10.1016/j.bone.2014.10.013
31. Tahir NT, Ahmed HS, Mohsen FY (2020) Vitamin D and risk of osteoarthritis among Iraqi patients with and without metabolic syndrome. Plant Archiv 20(1): 2731–2736
32. Beckett EL, Duesing K, Martin C et al (2016) Relationship between methylation status of vitamin D-related genes, vitamin D levels, and methyl-donor biochemistry. J Nutr Intermed Metab 6:8–15. https://doi.org/10.1016/j.jnim.2016.04.010
33. Haussler MR, Jurutka PW, Mizwicki M et al (2011) Vitamin D receptor (VDR)-mediated actions of 1α,25 $(OH)_2$ vitamin D3: genomic and non-genomic mechanisms. Best Pract Res Clin Endocrinol Metab 25(4):543–559. https://doi.org/10.1016/j.beem.2011.05.010
34. Nurminen V, Neme A, Seuter S et al (2018) The impact of the vitamin D-modulated epigenome on VDR target gene regulation. Biochim Biophys Acta (BBA)-Gene Regul Mech 1861(8):697–705. https://doi.org/10.1016/j.bbagrm.2018.05.006
35. Charoenngam N, Shirvani A, Kalajian TA et al (2020) The effect of various doses of oral vitamin D3 supplementation on gut microbiota in healthy adults: a randomized, double-blinded, dose-response study. Anticancer Res 40(1):551–556. https://doi.org/10.21873/anticanres.13984
36. Arora J, Wang J, Weaver V et al (2022) Novel insight into the role of the vitamin D receptor in the development and function of the immune system. J Steroid Biochem Mol Biol 219:106084. https://doi.org/10.1016/j.jsbmb.2022.106084
37. Ahmed HS, Ahmed HS (2021) Diabetes and covid-19 pandemic: a potential mechanisms: a review. IJFMT 15(1):2224–2231. https://doi.org/10.37506/ijfmt.v15i1.13734
38. Oh S, Chun S, Hwang S et al (2021) Vitamin D and exercise are major determinants of natural killer cell activity, which is age-and gender-specific. Front Immunol 12: 594356. https://doi.org/10.3389/fimmu.2021.594356
39. Carlberg C, Velleuer E (2022) Vitamin D and the risk for cancer: a molecular analysis. Biochem Pharmacol 196:114735. https://doi.org/10.1016/j.bcp.2021.114735
40. Manson JE, Bassuk SS, Buring JE (2020) Principal results of the vitamin D and OmegA-3 TriaL (VITAL) and updated meta-analyses of relevant vitamin D trials. J Steroid Biochem Mol Biol 198: 105522. https://doi.org/10.1016/j.jsbmb.2019.105522
41. Feige J, Moser T, Bieler L et al (2020) Vitamin D supplementation in multiple sclerosis: a critical analysis of potentials and threats. Nutrients 12(3):783. https://doi.org/10.3390/nu12030783
42. Manousaki D, Harroud A, Mitchell RE et al (2021) Vitamin D levels and risk of type 1 diabetes: a Mendelian randomization study. PLoS Med 18(2): e1003536. https://doi.org/10.1371/journal.pmed.1003536
43. Walls AC, Park Y-J, Tortorici MA et al (2020) Structure, function, and antigenicity of the sars-cov-2 spike glycoprotein. Cell 181(2):281–292.e6. https://doi.org/10.1016/j.cell.2020.02.058
44. Ahmed HS, Ahmed HS, Abud HN (2024) The role of vitamin D against covid-19 infection, progression

and severity. Hum Antibodies 32(2):51–60. https://doi.org/10.3233/HAB-240009
45. Mirea A-M, Pop RM, Căinap SS et al (2020) Presymptomatic diagnosis of CYP24A1-related infantile idiopathic hypercalcemia: a case report. Eur J Med Genet 63(12):104100. https://doi.org/10.1016/j.ejmg.2020.104100
46. Thakker R (2020) The parathyroid glands, hypercalcemia, and hypocalcemia. In: Schafer LGAI (ed) Goldman-Cecil medicine, 26th edn. Elsevier, Philadelphia,pp 1611–1623.e2
47. Sharma A, Thakker RV, Jüppner H (2009) Genetic disorders of calcium and phosphate homeostasis. In: Avner E, Harmon W, Niaudet P, Yoshikawa N (eds) Pediatric nephrology. Springer, Berlin, Heidelberg. https://doi.org/10.1007/978-3-540-76341-3_11
48. Waller D, Sampson A (2017) Calcium metabolism and metabolic bone disease. In: Medical pharmacology and therapeutics. Elsevier, Philadelphia, pp 481–489
49. Ahmed HS (2017) Correlations between serum interleukins-2,-4 levels and some biochemical parameters in Iraqi patients with osteoporosis. J Fac Med Baghdad 59(3):275–279. https://iqjmc.uobaghdad.edu.iq/index.php/19JFacMedBaghdad36/article/view/103
50. Raimann A, Mehany S, Feil P et al (2020) Decreased compressional sound velocity is an indicator for compromised bone stiffness in X-linked hypophosphatemic rickets (XLH). Front Endocrinol 11:355. https://doi.org/10.3389/fendo.2020.00355
51. Sıklar Z, Turan S, Bereket A et al (2020) Nationwide Turkish cohort study of hypophosphatemic rickets. J Clin Res Pediatr Endocrinol 12(2):150–159. https://doi.org/10.4274/jcrpe.galenos.2019.2019.0098
52. Ahmed HS, Ahmed HS, Ad'hiah AH (2020) Interleukin-1 single nucleotide polymorphisms and risk of systemic lupus erythematosus among Iraqi patients. Meta Gene 23:100640. https://doi.org/10.1016/j.mgene.2019.100640
53. Walker M, Shane E (2022) Hypercalcemia: a review. JAMA 328(16):1624–1636. https://doi.org/10.1001/jama.2022.18331
54. Jain N, Reilly R (2017) Hungry bone syndrome. Curr Opin Nephrol Hypertens 26(4):250–255. https://doi.org/10.1097/MNH.0000000000000327
55. Yamaguchi S, Hamano T, Doi Y et al (2020) Hidden hypocalcemia as a risk factor for cardiovascular events and all-cause mortality among patients undergoing incident hemodialysis. Sci Rep 10:4418. https://doi.org/10.1038/s41598-020-61459-4
56. Lightwood R, Stapleton T (1953) Idiopathic hypercalcaemia in infants. Lancet 265(6779):255–256. https://doi.org/10.1016/s0140-6736(53)90187-1
57. Schlingmann K, Kaufmann M, Weber S et al (2011) Mutations in CYP24A1 and idiopathic infantile hypercalcemia. N Engl J Med 365(5):410–421. https://doi.org/10.1056/NEJMoa1103864
58. Cavalier E, Huyghebaert L, Rousselle O et al (2020) Simultaneous measurement of 25(OH)-vitamin D and 24,25(OH)$_2$-vitamin D to define cut-offs for CYP24A1 mutation and vitamin D deficiency in a population of 1,200 young subjects. Clin Chem Lab Med 58(2):197–201. https://doi.org/10.1515/cclm-2019-0996
59. Miller A, Mathew S, Patel S et al (2022) Genetic disorders of calcium and phosphorus metabolism. Endocrine 3(1):150–167. https://doi.org/10.3390/endocrines3010014
60. Schlingmann K, Ruminska J, Kaufmann M et al (2016) Autosomal-recessive mutations in *SLC34A1* encoding sodium-phosphate cotransporter 2A cause idiopathic infantile hypercalcemia. J Am Soc Nephrol 27(2):604–614. https://doi.org/10.1681/ASN.2014101025
61. Pilz S, Theiler-Schwetz V, Pludowski P et al (2022) Hypercalcemia in pregnancy due to *cyp24a1* mutations: case report and review of the literature. Nutrients 14(12):2518. https://doi.org/10.3390/nu14122518
62. St-Arnaud R, Arabian A, Kavame D et al (2022) Vitamin D and diseases of mineral homeostasis: a *cyp24a1* r396w humanized preclinical model of infantilehypercalcemia type 1. Nutrients 14(15):3221. https://doi.org/10.3390/nu14153221
63. Bizerea-Moga TO, Chisavu F, Ilies C et al (2023) Phenotype of idiopathic infantile hypercalcemia associated with the heterozygous pathogenic variant of *SLC34A1* and *CYP24A1*. Children 10(10):1701. https://doi.org/10.3390/children10101701
64. Hureaux M, Chantot-Bastaraud S, Cassinari K et al (2021) When a maternal heterozygous mutation of the CYP24A1 gene leads to infantile hypercalcemia through a maternal uniparental disomy of chromosome 20. Mol Cytogenet 14(1):23. https://doi.org/10.1186/s13039-021-00543-4
65. Taylor-Miller T, Allgrove J (2021) Endocrine diseases of newborn: epidemiology, pathogenesis, therapeutic options, and outcome "current insights into disorders of calcium and phosphate in the newborn". Front Pediatr 9:600490. https://doi.org/10.3389/fped.2021.600490
66. Uwitonze AM, Rahman S, Ojeh N et al (2020) Oral manifestations of magnesium and vitamin D inadequacy. J Steroid Biochem Mol Biol 200:105636. https://doi.org/10.1016/j.jsbmb.2020.105636
67. Goodson JM, Shi P, Mumena CH et al (2017) Dietary phosphorus burden increases cariogenesis independent of vitamin D uptake. J Steroid Biochem Mol Biol 167:33–38. https://doi.org/10.1016/j.jsbmb.2016.10.006
68. Rausch S, Foller M (2022) The regulation of FGF23 under physiological and pathophysiological conditions. Pflugers Arch 474(3):281–292. https://doi.org/10.1007/s00424-022-02668-w
69. Simic P, Babitt JL (2021) Regulation of FGF23: beyond bone. Curr Osteoporos Rep 19(6):563–573. https://doi.org/10.1007/s11914-021-00703-w
70. Fukumoto S (2021) FGF23-related hypophosphatemic rickets/ osteomalacia: diagnosis

and new treatment. J Mol Endocrinol 66(2):R57–R65. https://doi.org/10.1530/JME-20-0089
71. Hetz R, Beeler E, Janoczkin A et al (2022) Excessive inorganic phosphate burden perturbed intracellular signaling: quantitative proteomics and phosphoproteomics analyses. Front Nutr 8:765391. https://doi.org/10.3389/fnut.2021.765391
72. Bilezikian J, Bandeira L, Khan A et al (2018) Hyperparathyroidism. Lancet 391(10116):168–178. https://doi.org/10.1016/S0140-6736(17)31430-7
73. Robinson C (2020) Genetics of mineral disorders. In: Zaidi M (ed) Encyclopedia of bone biology. Academic Press, Oxford, UK, pp 92–107
74. Diao J, DeBono A, Josephs T et al (2021) Therapeutic opportunities of targeting allosteric binding sites on the calcium-sensing receptor. ACS Pharmacol Transl Sci 4(2):666–679. https://doi.org/10.1021/acsptsci.1c00046
75. Warner J, Nyholt D, Busfield F et al (2006) Familial isolated hyperparathyroidism is linked to a 1.7 Mb region on chromosome 2p13.3–14. J Med Genet 43(3):e12. https://doi.org/10.1136/jmg.2005.035766
76. Shibata Y, Yamazaki M, Takei M et al (2015) Early-onset, severe, and recurrent primary hyperparathyroidism associated with a novel CDC73 mutation. Endocr J 62(7):627–632. https://doi.org/10.1507/endocrj.EJ15-0057
77. Simonds W, James-Newton L, Agarwal S et al (2002) Familial isolated hyperparathyroidism: clinical and genetic characteristics of 36 kindreds. Medicine 81(1):1–26. https://doi.org/10.1097/00005792-200201000-00001
78. Yang D, Zheng J, Tang F et al (2022) A two-generation hyperparathyroidism-jaw tumor (HPT-JT) syndrome family: clinical presentations, pathological characteristics and genetic analysis: a case report. Diagn Pathol 17(1):71. https://doi.org/10.1186/s13000-022-01248-x
79. Ibrahem HM (2020) Ossifying fibroma of the jaw bones in hyperparathyroidism-jaw tumor syndrome: analysis of 24 cases retrieved from literatures. J Dent Sci 15(4):426–432. https://doi.org/10.1016/j.jds.2019.12.007
80. Mele C, Mencarelli M, Caputo M et al (2020) Phenotypes associated with MEN1 syndrome: A focus on genotype-phenotype correlations. Front Endocrinol 11:591501. https://doi.org/10.3389/fendo.2020.591501
81. Murray SL, Wolf M (2024) Calcium and phosphate disorders: Core curriculum 2024. AJKD 83(2):241–256. https://doi.org/10.1053/j.ajkd.2023.04.017
82. Roszko K, Bi R, Mannstadt M (2016) Autosomal dominant hypocalcemia (hypoparathyroidism) types 1 and 2. Front Physiol 7:458. https://doi.org/10.3389/fphys.2016.00458
83. Gordon R, Levine M (2018) Genetic disorders of parathyroid development and function. Endocrinol Metab Clin N Am 47(4):809–823. https://doi.org/10.1016/j.ecl.2018.07.007
84. Ahmed HS (2017) A comparative estimation of metabolic and hormonal parameters among Iraqi hypothyroid patients. IPMJ 16(2):204–209
85. Sun M, Wu X, Yu Y et al (2020) Disorders of calcium and phosphorus metabolism and the proteomics/metabolomics-based research.Front Cell Dev Biol 8(576110):1–13. https://doi.org/10.3389/fcell.2020.576110
86. Thiele S, de Sanctis L, Werner R et al (2011) Functional characterization of GNAS mutations found in patients with pseudohypoparathyroidism type Ic defines a new subgroup of pseudohypoparathyroidism affecting selectively Gs-α-receptor interaction. Hum Mutat 32(6):653–660. https://doi.org/10.1002/humu.21489
87. Vlachopapadopoulou E-A, Anagnostou E, Dikaiakou E et al (2020) Pseudohypoparathyroidism type 1B (PHP1B), a rare disorder encountered in adolescence. J Pediatr Endocrinol Metab 33(11):1475–1479. https://doi.org/10.1515/jpem-2020-0192
88. Institute of Medicine (US) Standing Committee on the Scientific Evaluation of Dietary Reference Intakes (1997) Dietary reference intakes for calcium, phosphorus, magnesium, vitamin D, and fluoride. National Academies Press (US), Washington, DC. https://doi.org/10.17226/5776
89. Linglart A, Menguy C, Couvineau A et al (2011) Recurrent PRKAR1A mutation in acrodysostosis with hormone resistance. N Engl J Med 364(23):2218–2226. https://doi.org/10.1056/NEJMoa1012717
90. Liamis G, Hoorn E, Florentin M et al (2021) An overview of diagnosis and management of drug-induced hypomagnesemia. Pharmacol Res Prespect 9(4):e00829. https://doi.org/10.1002/prp2.829
91. Malinowska J, Małecka M, Ciepiela O (2020) Variations in magnesium concentration are associated with increased mortality: study in an unselected population of hospitalized patients. Nutrients 12(6):1836. https://doi.org/10.3390/nu12061836
92. Wang N, Chen Y, Ji J et al (2020) The relationship between serum vitamin D and fracture risk in the elderly: a meta-analysis. J Orthop Surg Res 15(1):81. https://doi.org/10.1186/s13018-020-01603-y
93. Risco F, Traba M (1994) Possible involvement of a magnesium dependent mitochondrial alkaline phosphatase in the regulation of the 25-hydroxyvitamin D3-1alpha-and 25-hydroxyvitamin D3-24R-hydroxylases in LLC-PK1 cells. Magnes Res 7(3–4):169–178
94. Ashique S, Kumar S, Hussain A et al (2023) A narrative review on the role of magnesium in immune regulation, inflammation, infectious diseases, and cancer. J Health Popul Nutr 42(1):74. https://doi.org/10.1186/s41043-023-00423-0

95. DiNicolantonio J, O'Keefe J (2021) Magnesium and vitamin D deficiency as a potential cause of immune dysfunction, cytokine storm and disseminated intravascular coagulation in covid-19 patients. Mo Med 118(1):68–73. PMCID: PMC7861592
96. Lu W, Pringa E, Chou L (2017) Effect of magnesium on the osteogenesis of normal human osteoblasts. Magnes Res 30(2):42–52. https://doi.org/10.1684/mrh.2017.0422
97. Malik M, Latiwesh O, Nouh F et al (2020) Response of parathyroid hormone to vitamin D deficiency in otherwise healthy individuals. Cureus 12(8): e9764. https://doi.org/10.7759/cureus.9764
98. Moe SM (2008) Disorders involving calcium, phosphorus, and magnesium. Prim Care Clin Off Pract 35(2):215–237. https://doi.org/10.1016/j.pop.2008.01.007
99. Wessells K, Brown K (2012) Estimating the global prevalence of zinc deficiency: results based on zinc availability in national food supplies and the prevalence of stunting. PLoS One 7(11):e50568. https://doi.org/10.1371/journal.pone.0050568
100. Kambe T, Taylor K, Fu D (2021) Zinc transporters and their functional integration in mammalian cells. J Biol Chem 296:100320. https://doi.org/10.1016/j.jbc.2021.100320
101. Razzaque M (2020) COVID-19 pandemic: can maintaining optimal zinc balance enhance host resistance? Tohoku J Exp Med 251(3):175–181. https://doi.org/10.1620/tjem.251.175
102. Akimbekov N, Digel I, Sherelkhan D et al (2020) Vitamin D and the host-gut microbiome: a brief overview. Acta Histochem Cytoc 53(3):33–42. https://doi.org/10.1267/ahc.20011
103. Matthews J, Sunde M (2002) Zinc fingers–folds for many occasions. IUBMB Life 54(3):351–355. https://doi.org/10.1080/15216540216035
104. Ahmed HS (2017) Metabolic and hormonal changes associated with menopause. MMJ 16(3):77–82
105. Mechica J, Leite M, Mendonca B et al (1997) A novel nonsense mutation in the first zinc finger of the vitamin D receptor causing hereditary 1,25-dihydroxyvitamin D3-resistant rickets. J Clin Endocrinol Metab 82(11):3892–3894. https://doi.org/10.1210/jcem.82.11.4384
106. Zhu W, Malloy P, Delvin E et al (1998) Hereditary 1,25-dihydroxyvitamin D-resistant rickets due to an opal mutation causing premature termination of the vitamin D receptor. J Bone Miner Res 13(2): 259–264. https://doi.org/10.1359/jbmr.1998.13.2.259
107. Pang Q, Qi X, Jiang Y et al (2016) Clinical and genetic findings in a Chinese family with VDR-associated hereditary vitamin D-resistant rickets. Bone Res 4(16018):1–7. https://doi.org/10.1038/boneres.2016.18
108. Amos A, Razzaque M (2022) Zinc and its role in vitamin D function. Curr Res Physiol 5:203–207. https://doi.org/10.1016/j.crphys.2022.04.001
109. Willems BA, Vermeer C, Reutelingsperger CP et al (2014) The realm of vitamin K dependent proteins: shifting from coagulation toward calcification. Mol Nutr Food Res 58(8):1620–1635. https://doi.org/10.1002/mnfr.201300743
110. Zoch ML, Clemens TL, Riddle RC (2016) New insights into the biology of osteocalcin. Bone 82: 42–49. https://doi.org/10.1016/j.bone.2015.05.046
111. Barrett H, O'Keeffe M, Kavanagh E et al (2018) Is matrix gla protein associated with vascular calcification? A systematic review. Nutrients 10(4):415. https://doi.org/10.3390/nu10040415
112. Nigwekar SU, Bloch DB, Nazarian RM et al (2017) Vitamin K-dependent carboxylation of matrix gla protein influences the risk of calciphylaxis. J Am Soc Nephrol 28(6):1717–1722. https://doi.org/10.1681/ASN.2016060651
113. Mohammed MS, Ahmed HS (2023) Urokinase-type plasminogen activator receptor as a predictive marker for cardiac disease among type 2 diabetic patients. MJB 20(4):750–756. https://doi.org/10.4103/MJBL.MJBL_832_23
114. Seefried L, Kishnani PS, Moseley S et al (2021) Pharmacodynamics of asfotase alfa in adults with pediatric-onset hypophosphatasia. Bone 142: 115664. https://doi.org/10.1016/j.bone.2020.115664
115. Zittermann A, Pilz S, Berthold HK (2020) Serum 25-hydroxyvitamin D response to vitamin D supplementation in infants: a systematic review and meta-analysis of clinical intervention trials. Eur J Nutr 59(1):359–369. https://doi.org/10.1007/s00394-019-01912-x
116. Irzyniec T, Bory'n M, Kasztalska J et al (2020) The effect of an oral sodium phosphate load on parathyroid hormone and fibroblast growth factor 23 secretion in normo- and hypercalciuric stone-forming patients. Clin Nutr 39(12):3804–3812. https://doi.org/10.1016/j.clnu.2020.04.020

Vitamin D and Sleep Disorders

Sadeka Tamanna

Abstract

Sleep is a complex behavioral state necessary for neurological, physical, cognitive, metabolic, and cardiovascular health throughout the lifespan. Poor sleep increases the risk of mortality among the general population. Sleep deprivation causes memory deficits, neuropsychiatric diseases, mood disorders, metabolic syndrome, cardiovascular diseases, stroke, and neurodegenerative diseases and carries an increased risk of suicide and motor vehicle accidents. Vitamin D is a fat-soluble secosteroid and its inadequacy is associated with multiple disorders, including bone health, neuropsychiatric diseases, metabolic disorders, and cardiovascular disorders. In recent years, vitamin D deficiency has been shown to affect sleep quality and many sleep disorders. Some studies have shown improvement in sleep disorders with vitamin D replacement, whereas others have not found any significant change. This chapter elaborates the role of vitamin D in different sleep disorders and describes the potential suggested mechanisms.

Keywords

Sleep · Apnea · Insomnia · Vitamin D

Introduction

Sleep is a state of reduced responsiveness to the environment when the brain is active in a different mode, restoring, repairing, and increasing the functions of the body. Sleep is not a mere absence of wakefulness. Approximately one-third of the lifespan is spent in sleep for this purpose, and sleep is crucial to the survival of the animal. There are two types of sleep, non-rapid eye movement sleep (NERM) and rapid eye movement sleep (REM). Each cycle of sleep lasts approximately 90–120 minutes, where NREM sleep comprises 75–80% and REM sleep comprises approximately 25–20%. The initiation and maintenance of sleep require the suppression of the arousal system and the inhibition of wakefulness-promoting neurotransmitters. Severe sleep restriction has led to premature death among animals due to the accumulation of reactive oxygen species [1, 2]. Multiple epidemiological studies have also shown that sleeping <7 hours or >8 hours per night increases all-cause mortality [3, 4]. Overall, sleep deprivation causes cognitive and memory deficits, neuropsychiatric diseases, mood disorders, metabolic syndrome, cardiovascular diseases, stroke, the progression of neurodegenerative diseases, and an increased risk of suicide

S. Tamanna (✉)
Department of Internal Medicine and Sleep Medicine, University of Mississippi Medical Center, Jackson, MS, USA

G. V. (Sonny) Montgomery VA Medical Center, Jackson, MS, USA
e-mail: sadeka.tamanna@va.gov; stamanna@umc.edu

M. S. Razzaque (ed.), *Vitamin D Function*, Advances in Experimental Medicine and Biology 1493,
https://doi.org/10.1007/978-3-032-04357-3_7

and motor vehicle accidents [5]. Poor sleep quality affects daytime alertness and decreased responsiveness to important tasks. Sleep disorders, including obstructive sleep apnea, insomnia, restless leg syndrome, and narcolepsy, negatively impact quality of life.

Vitamin D is a pleiotropic hormone with both skeletal and nonskeletal functions. Vitamin D exists in two forms in nature: D3 and D2. Vitamin D3 (cholecalciferol) is produced in the skin from 7-hydroxy cholesterol by ultraviolet light. D2 (ergocalciferol) comes from plants and dietary sources. The serum plasma level of 25 (OH) vitamin D is usually used to assess vitamin D status since it reflects contribution from both diet and synthesis in the skin [6]. The estimated prevalence of vitamin D deficiency (25 (OH) vitamin D < 30 nm/L) globally is 15.7%, according to a pooled analysis of 7.9 million people worldwide [7]. There has been studies linking vitamin D deficiency and sleep in recent years. Many clinical studies have indicated that vitamin D deficiency is associated with poor sleep quality and sleep disorders. Some studies also revealed a low level of vitamin D among patients diagnosed with obstructive sleep apnea (OSA), where the treatment of OSA returned the vitamin D level to normal. A multiethnic study revealed a strong association between vitamin D deficiency and short sleep duration among African Americans with a high apnea-hypopnea index and obstructive sleep apnea. Vitamin D acts after it binds to the vitamin D receptor (VDR) on cells [8]. The presence of VDR in sleep-promoting areas of the brain has been reported, which also helps elucidate the relationship between vitamin D deficiency and sleep regulation. The purpose of this chapter is to review the physiological relationship between vitamin D and sleep and the effects of vitamin D deficiency on different sleep disorders.

Vitamin D

As mentioned, there are two forms of vitamin D—D3 (cholecalciferol) and D2 (ergocalciferol)—both undergo hydroxylation at the liver to 25(OH) vitamin D. This 25 (OH) vitamin D then moves to the kidney for further hydroxylation by 1alpha hydroxylase enzyme (CYP27B1) to generate biologically active 1,25(OH)vitamin D or calcitriol. However, other tissues, such as those of the central nervous system, also express CYP27B1 and possess the capacity to make calcitriol independently. Vitamin D and its metabolites are carried in blood bound to vitamin D binding protein (DBP). Vitamin D exerts biological effects in both genomic and non-genomic ways (Fig. 1). Classic genomic methods work through the nuclear VDR, which takes hours to days. The VDR is present not only in these organs but also in most cells of the body, including the brain. Immunohistochemical staining revealed its presence in the hypothalamus, supraoptic nuclei, and dopaminergic neurons in the substantia nigra which are involved in sleep regulation [9]. In this process, calcitriol binds to the VDR, a transcription factor in the cell nucleus, and heterodimerizes with the retinoid X receptor (RXR). This calcitriol-VDR-RXR complex couples to a specific sequence of DNA known as the vitamin D response element (VDRE) [10]. This VDRE exerts its effects through several target genes and coactivator molecules to mediate its biological activity; non-genomic action is a rapid, non-transcriptional effect mediated by membrane-bound receptors and signaling pathways. It works by opening or ion channels, second messenger system signaling, activation of intracellular signaling molecules, calcium influx within the cell, and modulation of downstream pathways [11]. The VDR is not restricted to musculoskeletal tissues; rather, it is widely distributed throughout the body. The VDR is widely present in the brain, including the pituitary, cerebral cortex, cerebellum, amygdala, anterior cingulate gyrus, hippocampus, hypothalamus, putamen, and spinal cord. Vitamin D has genomic and non-genomic functions to include calcium transport across the plasma membrane [12].

Vitamin D receptors are present in brain areas involved in sleep regulation, such as the

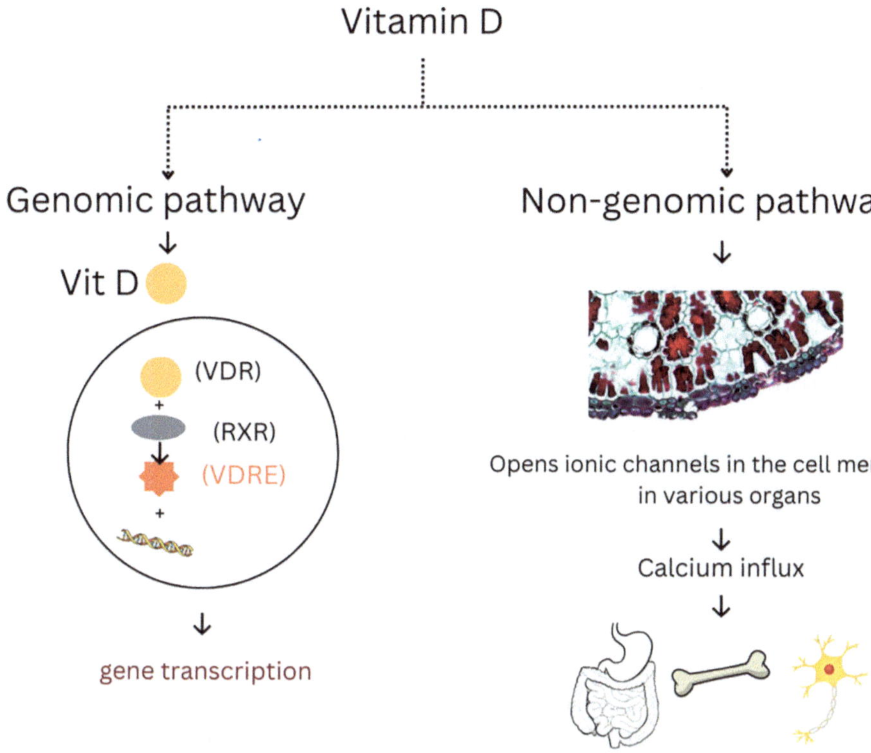

Fig. 1 Mechanism of action of vitamin D at the cellular level

hypothalamus and the pineal gland. The effects of vitamin D on sleep are multifaceted and primarily involve neuroendocrine pathways, anti-inflammatory effects, and circadian rhythm regulation (Fig. 2). The sleep-wake cycle is operated by groups of sleep-promoting and wake-promoting neurons that secrete specific neurotransmitters and peptides in a hierarchical manner. Dopamine, histamine, serotonin, norepinephrine, hypocretin, and neuropeptides are wake-promoting neurotransmitters, whereas melanin-concentrating hormones, adenosine, gamma amino butyric acid (GABA), and melatonin are sleep-promoting neurotransmitters.

Serotonin-Melatonin Pathway Regulation

Tryptophan to Serotonin Conversion Vitamin D increases the level of the enzyme tryptophan hydroxylase (TPH), which converts tryptophan to serotonin. Serotonin, a potent wake-promoting neurotransmitter, functions to promote waking and inhibit REM sleep. Serotonin-containing cells of the dorsal raphe nucleus innervate other areas of the brain to promote wakefulness. This occurs in both peripheral tissues and the brain, though serotonin synthesis in the brain is compartmentalized through the blood-brain barrier [13, 14].

Melatonin Synthesis Serotonin is a precursor for melatonin, the hormone critical for sleep-wake cycles. Vitamin D may enhance melatonin production indirectly by increasing serotonin availability, particularly in the pineal gland, where melatonin is synthesized in response to darkness.

Neurotransmitter Systems

GABA Vitamin D may enhance GABA synthesis or receptor function, promoting relaxation and

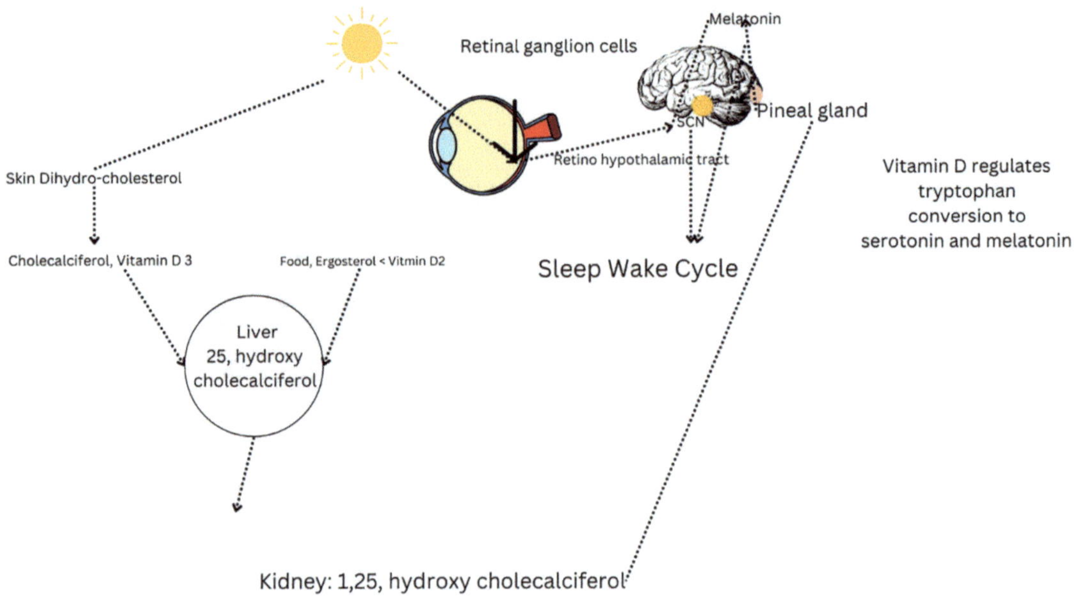

Fig. 2 Role of vitamin D in the regulation of sleep

sleep. Animal studies suggest that vitamin D influences GABA pathways, although evidence in humans is limited. Vitamin D affects the signaling system of glutamate and GABA, which are major excitatory and inhibitory neurotransmitters. A vitamin D-deficient diet fed to rodents resulted in low glutamate and glutamine levels and high GABA levels in the brain [15]. Dysfunction of the GABA-ergic system can cause sleep disruption, hypersomnia, and restless leg syndrome. Vitamin D deficiency in rodent models results in altered behavior, decreased levels of glutamate and glutamine, and increased levels of GABA and glycine [15].

Dopamine Another important awake neurotransmitter is also affected by vitamin D deficiency. Prenatal deficiency of vitamin D alters the dopaminergic pathway in the adult brain, disrupting behavior and neurochemistry in rodents [16]. The nuclear location of VDR in dopaminergic neurons suggests that vitamin D may regulate cell proliferation, differentiation, and survival [16]. Deficiency in prenatal vitamin D has been shown to influence dopaminergic activity in adult life.

Circadian Rhythm and Suprachiasmatic Nucleus (SCN) Interaction

SCN Receptors The SCN, the brain's master circadian clock, contains VDRs. Vitamin D may influence circadian gene expression (e.g., CLOCK and PER genes), helping synchronize sleep-wake cycles. It receives input from the retinal cells of the eyes with light exposure. As light is absorbed by the eyes, it triggers SCN to establish the body's central clock to a rhythm of approximately 24 hours, primarily entraining due to light-darkness cycles and forcing circadian rhythms of physiology and behavior [17].

Light Exposure Synergy Exposure to sunlight increases vitamin D production and controls the circadian rhythms. Morning sunlight may improve both the levels of vitamin D and SCN activity, inducing alertness during the day and melatonin secretion at night.

Immunoinflammatory Role Vitamin D helps in the regulation of T and B cells, dendritic cells, and macrophages and in inhibiting the secretion of pro-cytokines [18]. Vitamin D suppresses inflammatory cytokines (e.g., TNF-α and IL-6) linked with sleep disturbance. Insomnia and OSA are both related to long-term inflammation. Reducing inflammation, vitamin D can improve sleep.

Indirect Effects via Comorbidities

Mood and Pain Vitamin D deficiency is linked to depression and chronic pain, both of which impair sleep. Correcting deficiency may alleviate these conditions indirectly. Although a clear pathway of sleep disruption through vitamin D deficiency has not yet been established, clinical studies have shown that vitamin D deficiency increases insomnia, sleep disruption, daytime somnolence, and restless leg syndrome.

Vitamin D Deficiency and Sleep Disorders

Poor sleep quality adversely affects overall health. The Pittsburg Sleep Quality Index (PSQI) is the most widely used subjective tool to understand and gauge the level of sleep [19]. Actigraphy, overnight polysomnography, or home sleep studies can objectively measure sleep quality by time of wake after sleep onset (WASO), sleep efficiency (SE), and other standardized measures according to the device used. Poor sleep quality can result from complete or partial sleep deprivation, a change in sleep architecture with selective sleep stage deprivation, or sleep restriction from sleep debt [5]. Studies have proven that individuals with low levels of vitamin D are more prone to poor quality sleep and reduced sleep duration. For example, a meta-analysis established that a deficiency in vitamin D was positively associated with an enhanced risk of sleep disorders, poor sleep quality, and short sleep duration. This meta-analysis included 9 studies with 9397 participants and concluded that vitamin D deficiency had a strong association with high risk of sleep disorders, poor quality of sleep, brief sleep duration, and sleepiness, with odds ratio for sleep disorders of 1.50 (95% CI: 1.31, 1.72). Brief sleep duration of <7 hours has been related to negative health outcomes and increased mortality [20]. Micronutrient deficiency, including vitamin D deficiency, has been shown to reduce sleep duration in an analysis of NHANES data from 2005 to 2016 [21]. A cross-sectional study of 3048 community dwellers aged 65 years and older revealed that low vitamin D was associated with short sleep time, poor sleep efficiency, and wake time after sleep onset (WASO) in actigraphy data worn for 24 hours for 5 days [22]. Vitamin D deficiency in children who underwent overnight polysomnography was associated with decreased sleep duration and poor sleep efficiency [23]. In patients with stroke, low vitamin D levels and pre-stroke poor sleep quality, as measured with the PSQI, increase the risk of poststroke depression and anxiety [24, 25]. Level of 25(OH) vitamin D level was assessed in 81 sleep clinic patients who complained of sleep problems and nonspecific pain. Excessive sleepiness was measured with the EPWORTH sleepiness score (ESS) among them. ESS was found to be inversely correlated with the level of vitamin D when the level of vitamin D was >20 ng/ml. ESS was directly correlated with vitamin D levels when the level was <20 ng/ml among black patients but not in white patients. The mean ESS was higher among black patients when the vitamin D concentration was less than 20 ng/ml. These findings indicate that there is a difference in the effect of low vitamin D concentrations among blacks and white individuals [26]. This deficiency may have impacted the rise of sleep disorders over the last few decades, but more understanding through clinical and translational research is needed.

Working indoors and spending most of the day inside deprived people of sunlight, which was not an issue about a century ago. A cross-sectional study of 1472 subjects [27] working a fixed day in Korea revealed a significant association between low vitamin D levels and poor sleep quality. Most of the world population who work indoor for about 8–10 hours per day has reduced their sun exposure and vitamin D levels. A systematic review and meta-analysis of intervention studies found that vitamin D supplementation improved sleep quality in some cases, but the results were not consistent across all studies [28]. Standardized vitamin D supplementation for Veterans with insufficient vitamin D (20–29 ng/ml) or a deficient range (<20 ng/ml) was provided with vitamin D supplements, followed by their pre- and post-pain scores, sleep-related PSQI scores, and quality-of-life scores. The improvements were significant for sleep latency ($p = 0.019$), sleep duration ($p = 0.012$), pain score ($p = 0.001$), and general health ($p = 0.006$) [29]. A randomized controlled trial of vitamin D supplementation among patients with sleep disorders based on the PSQI revealed a significant improvement in sleep quality ($P < 0.05$) among those receiving vitamin D supplementation [30]. Another prospective study of 152 military soldiers showed a significant improvement in fatigue and sleep-related impairment ($p < 0.01$) among the treatment group receiving vitamin D supplements [31].

Obstructive Sleep Apnea (OSA)

Complete or partial collapse of the upper airway with desaturation of oxygen causing sleep disruption and excessive daytime sleepiness is the primary characteristic of OSA. It carries a prevalence of 9–38% [32]. Overnight polysomnography in a laboratory sleep study or home sleep studies are the two diagnostic measures for obstructive sleep apnea. The total number of apnea (complete cessation of breath) and hypopnea (partial reduction in breathing with desaturation of oxygen) events is considered over the total number of hours of sleep to measure the apnea-hypopnea index (AHI), which is used as a standardized measure of the severity of OSA. Snoring, waking up with difficulty breathing, excessive daytime sleepiness, headaches, poor cognition, concentration and memory difficulty, depression, and mood disorders are common symptoms of OSA. It also increases the risk of hypertension, cardiovascular diseases, stroke, and overall mortality. Collapsibility of the upper airway muscles to interact during sleep is the key factor in the development of sleep apnea [33]. Low vitamin D levels are associated with poor musculoskeletal function. It has been hypothesized that low vitamin D may play a role in the poor tone of the pharyngeal muscles necessary for keeping the airway open during sleep. This low tone of muscles may cause obstructive sleep apnea among patients with low vitamin D. The role of vitamin D has been investigated in several cross-sectional, randomized, controlled observational studies. A meta-analysis by Upala et al. revealed that patients with obstructive sleep apnea had lower vitamin D levels than controls did, with a total of 506 patients and a mean difference of −5.81 ($P = 0.0008$) in vitamin D levels [34]. Goswami et al. performed a cross-sectional study among 2827 community-dwelling elderly males. The participants underwent in-home polysomnography, and their vitamin D levels were checked. Those in the lowest quartile (6–23 ng/ml) had greater odds of developing severe sleep apnea (odds ratio = 1.45, 95% CI =1.02–2.07) (35–85 ng/ml) compared with those in the highest quartile. There was no evidence of an independent association between lower vitamin D levels and increased odds of mild and moderate obstructive sleep apnea. The association between low vitamin D and sleep apnea was confounded by body mass index and large neck size in this study [35]. A cross-sectional study of 217 patients with obstructive sleep apnea revealed that excessive daytime sleepiness among newly diagnosed OSA patients was associated with low vitamin D levels [36]. Serum vitamin D was found to be inversely related to the severity of OSA [37]. Vitamin D levels were compared between patients diagnosed with OSA (AHI > 5) and those without OSA

(AHI < 5), revealing a significantly lower level of vitamin D among OSA patients [38]. However, some studies did not find a relationship between vitamin D deficiency and the severity of obstructive sleep apnea. A large proportion of patients referred for sleep studies had vitamin D deficiency, but there was no association between OSA diagnosis or the severity of OSA and vitamin D levels [39]. Immune system modulation, myopathy, and inflammation have been proposed as mechanisms by which vitamin D deficiency causes OSA. The prevalence of vitamin D deficiency is high in obesity, which is also a risk factor for OSA. Thus, vitamin D could have a confounding relationship between these two conditions. Vitamin D is a fat-soluble vitamin that is sequestered in adipose tissue. An obese individual contains a large amount of excess fat, which reduces its release of vitamin D into the circulation [40]. Obesity is also a strong risk factor for OSA, which can confound the risk between vitamin D and OSA. A reduction in the strength of the pharyngeal dilator muscle, which primarily causes collapsibility of the upper airway during sleep, has been proposed as a mechanism of OSA caused by vitamin D deficiency [41]. Intermittent hypoxia during sleep is a classic feature among patients with OSA. In patients with OSA, several indices of hypoxia, including average and minimum oxyhemoglobin saturation, time spent <90% of oxygen saturation, and ODI during sleep, have been associated with vitamin D levels in some studies but not all. A meta-analysis of 11 studies with 506 subjects revealed that participants with OSA had lower vitamin D levels than controls did (−5.81) (95% confidence interval [CI]: −10.09 to −1.53, $p = 0.008$) [34].

Continuous positive airway pressure (CPAP) therapy is the mainstay of treatment for OSA. There is a significant increase in vitamin D levels after 1 year of CPAP therapy in patients with OSA [42]. Liguori et al. studied a group of patients with severe OSA (AHI > 30) and controls without OSA. Serum vitamin D, calcium, and PTH levels were measured at baseline and after 1 week of CPAP therapy. The vitamin D level was lower and the PTH level was greater in patients with OSA than in controls at baseline. They also divided the OSA group into responders (residual AHI < 5) and nonresponders (residual AHI > 5). After 7 days of CPAP therapy, the vitamin D level increased in male OSA-R patients (responders), although the same increase in the vitamin D level was not detected in female OSA-R patients. Moreover, OSA nonresponders did not show a reduction in vitamin D levels after CPAP therapy [42]. Another study revealed a significant increase in vitamin D levels among OSA patients who were >70% compliant with CPAP therapy compared with the noncompliant group after 1 year of CPAP therapy [43]. Not all studies reported a positive effect of CPAP therapy on vitamin D levels. A randomized, sham-controlled trial of CPAP therapy in severe OSA patients revealed that compared with sham therapy, 12 weeks of CPAP therapy did not affect the levels of vitamin D or any bone turnover markers [44].

Insomnia

About one-third of people suffer from insomnia globally [32]. Insomnia is defined as difficulty falling asleep, staying asleep, or nonrestorative sleep causing daytime impairment or sleepiness for more than a month. The prevalence varies by the type of tools used and the population in which they are used. The insomnia sleep index (ISI) is one of the most commonly used screening tools [45, 46]. Grimnes et al. hypothesized that the worldwide epidemic of sleep disorders is due to vitamin D deficiency because people have worked indoors over the last few decades. Fatigue, cognitive dysfunction, poor vigilance in task performance, depression, anxiety, and difficulty maintaining social interactions are very common among patients with insomnia. Suicidal ideation is also closely related to insomnia. In a sample of 311 military personnel, insomnia symptoms were found to be unique predictors of suicide attempt longitudinally after adjusting for hopelessness and other depression symptoms [47]. The vitamin D level has been linked to daytime sleepiness, showing an inverse relationship with the vitamin D level in a cross-sectional study of 81 sleep patients in the clinic. For a vitamin D level of <20 ng/ml, ESS was directly correlated with vitamin D in black patients ($r = 0.48$, $p < 0.05$) but not in white patients. The vitamin D level was measured in 712 people in a large cross-sectional study who underwent overnight polysomnography and completed questionnaires on sleep. Among men, individuals with a sleep duration of less than 6 hours presented a twofold increase in low vitamin D levels (<20 ng/ml) [48]. A genetic causal relationship was examined via two-sample Mendelian randomization. Vitamin D was directly proportional to sleep duration and positively correlated with trouble falling or staying asleep [49].

A large cross-sectional study among the Arctic population revealed a strong association between vitamin D levels and sleep duration in both sexes [46]. Vitamin D is thought to affect the modulation of signaling pathways in the central nervous system. High-dose vitamin D was given to 940 adolescent girls for 9 weeks. Their cognitive ability, insomnia, and daytime somnolence improved [50]. In a randomized trial, vitamin D-fortified low-fat milk in the diet improved insomnia symptoms (ISI) in the treatment group compared with the control group [51]. The results of these studies have varied. A randomized trial was performed on 189 vitamin D-deficient participants, who were divided into 2 groups. One group was provided with vitamin D, and the other group was given a placebo. The patients were followed for 4 months, and their sleep data were compared. The researchers did not find any significant differences in insomnia or other sleep data between these two groups [52].

Restless Leg Syndrome (RLS)

RLS is a somatosensory network disorder characterized by an abnormal urge to move the legs, associated with unpleasant, crawling leg sensations, induced or exacerbated by rest, and relieved by movement, with diurnal fluctuations in symptoms worsening in the evening [53]. RLS can be idiopathic or related to pregnancy, iron deficiency or chronic renal failure, diabetes mellitus, or rheumatological disorders. A family history of RLS is also very common, suggesting autosomal dominant transmission with high penetrance. There has been substantial evidence of genetic investigations showing the associations of several loci on chromosomes 12q, 14q, 9p, 2q, 20p, and 16p. Recent advancements in brain imaging and dopamine agonist treatment-success have indicated that dopaminergic dysfunction in the central nervous system may be responsible for RLS. The prevalence of RLS is approximately 5–10% in the general population [54]. Many studies have reported low vitamin D levels among patients with RLS. A recent meta-analysis revealed a significantly lower level of vitamin D in 538 people with RLS than in 1588 people without RLS. One hundred fifty patients with RLS underwent EMG and serum assessment of vitamin D levels. The presence of RLS was significantly greater in the vitamin D-deficient group than in the vitamin D-deficient group [55]. In a population-based study of 201 subjects, the odds of developing RLS was 4.24 ($p < 0.001$) for those with vitamin D levels <50 nmol/L [56].

It is not completely clear how vitamin D helps RLS. Dopamine dysfunction or deficiency in the CNS has been proposed to have a role on RLS. Dopamine agonists improve RLS symptoms whereas dopamine antagonists make it worse. Vitamin D plays a major role in dopamine-producing neurons. Vitamin D has been identified as a key factor in dopaminergic neurogenesis and differentiation. Vitamin D plays an important role in dopaminergic development by increasing neurite growth, branching, presynaptic protein distribution, dopamine production, and functional release in various in vitro models [16]. Vitamin D is neuroprotective and anti-inflammatory for the CNS. Vitamin D3 crosses the blood-brain barrier and exerts a neuroprotective effect on a Parkinson's disease in animal model by inhibiting pro-inflammatory processes and upregulating anti-inflammatory processes [57]. Decreased dopamine in the stratal area of the brain was seen in a PET study suggesting dopamine dysfunction among patients with RLS compared with controls [58]. All these studies suggest a unified mechanism of enhancing and protecting the dopaminergic function of vitamin D, alleviating symptoms of RLS. RLS responds to GABAergic medications such as gabapentin and pregabalin. Vitamin D deficiency can reduce the mRNA expression of GABA receptors, decrease initial GABA uptake by nerve terminals, and increase the number of GABA-positive reactive astrocytes [59]. Vitamin D improved the severity of RLS symptoms among patients with primary RLS and vitamin D deficiency [60, 61].

Narcolepsy

Narcolepsy is a chronic sleep disorder characterized by excessive daytime sleepiness and symptoms related to dysfunction of rapid eye movement (REM) sleep. Narcolepsy is classified on the basis of the presence or absence of cataplexy or hypocretin deficiency. Narcolepsy type I (NT1) is defined by the presence of cataplexy and low hypocretin in CSF, and narcolepsy type II (NT2) is defined by the absence of cataplexy and normal hypocretin when measured. Loss of hypocretin results from the destruction of hypocretin-producing cells in the hypothalamus, which is thought to be autoimmune-mediated. Hypocretin stabilizes wakefulness and makes sleep transitions possible, and it also prevents the intrusion of REM sleep into wakefulness. Sleep-related hallucinations and sleep paralysis are also features of both types of narcolepsy [62].

Vitamin D deficiency has been linked to narcolepsy in recent studies. The vitamin D level was measured in 51 patients with narcolepsy with cataplexy (NC) and 55 controls in an age- and sex-matched group. The serum vitamin D level was lower in the NC group than in the control group ($P < 0.0039$) [18]. However, some studies did not find any difference. Vitamin D levels were measured in 174 Caucasian patients with NT1 and 174 controls. The vitamin D level did not differ between these two groups [63]. A case study of a 28-year-old female with hypersomnia revealed that her vitamin D level was very low at 5.9 ng/ml and that vitamin D replacement resolved her hypersomnia. Her initial sleep latency was 5.5 minutes in multiple sleep latency tests, which improved to 17 minutes after vitamin D replacement [25]. The role of vitamin D in narcolepsy is still not completely understood because of variable results.

Vitamin D Supplementation

Some studies have explored the effects of vitamin D supplementation on sleep. A systematic review and meta-analysis of intervention studies revealed that vitamin D supplementation improved sleep quality in some cases, but the results were not consistent across all studies [64]. Standardized vitamin D supplements for Veterans with insufficient vitamin D (20–29 ng/ml) or a deficient range (<20 ng/ml) were provided with vitamin D supplements, followed by their pre- and post-pain scores, sleep-related PSQI scores, and quality-of-life scores. The improvements were significant for sleep latency ($p = 0.019$), sleep duration ($p = 0.012$), pain score ($p = 0.001$), and general health ($p = 0.006$) [29]. A randomized controlled trial of vitamin D supplementation among patients with sleep disorders based on the PSQI revealed a significant improvement in sleep quality ($P < 0.05$) among those receiving vitamin D supplementation [30, 51]. Another prospective study of 152 military soldiers revealed a significant improvement in fatigue and sleep-related impairment ($p < 0.01$) among the treatment group receiving vitamin D supplements [31]. There are variable results concerning the effects of vitamin D in different studies. A pilot, double-blind RCT revealed no change in CPAP usage, BMI, or medication in patients with OSA who were given vitamin D supplements [65].

Clinical Implications

1. *Sleep disorders*: Vitamin D deficiency has been linked to all types of sleep disorders, such as insomnia, OSA, narcolepsy, and RLS. Correcting vitamin D deficiency will improve sleep quality and reduce the severity of these disorders.
2. *Overall health*: Poor sleep quality increases risk of various health problems like cardiovascular diseases, metabolic disorders, and

psychiatric diseases. Having adequate vitamin D levels can improve overall health by a promoting restorative sleep.

3. *Public health*: Given the widespread prevalence of vitamin D deficiency and its impact on sleep, addressing this will yield significant public health benefits.

Recommendations

1. *Vitamin D supplementation*: Supplementation may improve sleep quality in individuals with vitamin D deficiency [66]. For newer vitamin D guidelines for testing and dosing, readers are recommended to a recent publication [67] and chapter "Decoding Vitamin D Intake: How Much Is Too Much?" of this book [68].
2. *Sun exposure*: Encouraging safe sun exposure can help increase vitamin D levels. Spending 10–30 minutes in the sun several times a week, especially at midday, can be beneficial.
3. *Dietary intake*: Consuming foods rich in vitamin D, such as fatty fish, fortified dairy products, and egg yolks, can help maintain adequate levels.
4. *Regular monitoring*: Regular monitoring of vitamin D levels through blood tests can help identify deficiencies early and guide appropriate interventions.

Conclusion

There is a substantial correlation between vitamin D deficiency and poor sleep quality and some specific sleep disorders. The mechanism of action of vitamin D is not fully understood, but modulation of melatonin, regulation of circadian rhythm, and regulation of sleep-promoting neurotransmitters such as serotonergic and dopaminergic pathways have been elucidated. Correcting vitamin D deficiency through diet, supplementation, and safe sun exposure can significantly improve sleep quality and overall quality of health. Regular monitoring of vitamin D levels is recommended for early diagnosis and treatment. Longitudinal studies are required to understand the effects of vitamin D deficiency on various aspects of sleep disorders. More randomized controlled trials should be done to determine the effects of vitamin D on specific biological markers related to sleep disorders, sleep quality, and daytime sleepiness. Biological, mechanistic, and translational research is necessary to understand the pathways by which vitamin D affects the circadian rhythm. This information will help guide specific interventions for sleep disorders. Large population-based studies are important in determining the effect of vitamin D on different races, ethnic groups, demographics by sun distributions, and exposures to their sleep patterns. Research is also necessary to determine the specific dosages of vitamin D most effective for the treatment of sleep disorders in different groups of population. Public health campaigns aimed at increasing awareness and proper consumption of vitamin D can significantly influence sleep health and overall well-being. A meaningful step toward better health and quality of life can be achieved by addressing the link between vitamin D and sleep.

References

1. Brewer LD, Thibault V, Chen KC, Langub MC, Landfield PW, Porter NM (2001) Vitamin D hormone confers neuroprotection in parallel with downregulation of L-type calcium channel expression in hippocampal neurons. J Neurosci 21:98–108
2. Vierucci F, Fusani L, Saba A et al (2022) Gestational vitamin D3 supplementation and sun exposure significantly influence cord blood vitamin D status and 3-epi-25-hydroxyvitamin D3 levels in term newborns. Clin Chim Acta 524:59–68
3. Ikehara S, Iso H, Date C et al (2009) Association of sleep duration with mortality from cardiovascular disease and other causes for Japanese men and women: the JACC study. Sleep 32:295–301
4. Ayas NT, White DP, Manson JAE et al (2003) A prospective study of sleep duration and coronary heart disease in women. Arch Intern Med 163:205–209
5. Banks S, Dinges DF (2007) Behavioral and physiological consequences of sleep restriction. J Clin Sleep Med 3:519–528
6. Cashman KD, van den Heuvel EGHM, Schoemaker RJW, Prévéraud DP, Macdonald HM, Arcot J (2017)

25-Hydroxyvitamin D as a biomarker of Vitamin D status and its modeling to inform strategies for prevention of Vitamin D deficiency within the population. Adv Nutr 8:947–957
7. Cui A, Zhang T, Xiao P, Fan Z, Wang H, Zhuang Y (2023) Global and regional prevalence of vitamin D deficiency in population-based studies from 2000 to 2022: a pooled analysis of 7.9 million participants. Front Nutr 10:1070808
8. Bertisch SM, Sillau S, De Boer IH, Szklo M, Redline S (2015) 25-Hydroxyvitamin D concentration and sleep duration and continuity: multi-ethnic study of atherosclerosis. Sleep 38:1305–1311
9. Eyles DW, Smith S, Kinobe R, Hewison M, McGrath JJ (2005) Distribution of the Vitamin D receptor and 1α-hydroxylase in human brain. J Chem Neuroanat 29: 21–30
10. Muscogiuri G, Barrea L, Scannapieco M et al (2019) The lullaby of the sun: the role of vitamin D in sleep disturbance. Sleep Med 54:262–265
11. Bikle DD (2014) Vitamin D metabolism, mechanism of action, and clinical applications. Chem Biol 21: 319–329
12. Bikle DD, Vitamin D (2018) Biochemistry and physiology. Contemp Endocrinol:1–40
13. Patrick RP, Ames BN (2014) Vitamin D hormone regulates serotonin synthesis. Part 1: relevance for autism. FASEB J 28:2398–2413
14. Monti JM (2010) The role of dorsal raphe nucleus serotonergic and non-serotonergic neurons, and of their receptors, in regulating waking and rapid eye movement (REM) sleep. Sleep Med Rev 14:319–327
15. Groves NJ, Kesby JP, Eyles DW, McGrath JJ, Mackay-Sim A, Burne THJ (2013) Adult vitamin D deficiency leads to behavioural and brain neurochemical alterations in C57BL/6J and BALB/c mice. Behav Brain Res 241:120–131
16. Pertile RAN, Brigden R, Raman V, Cui X, Du Z, Eyles D (2023) Vitamin D: a potent regulator of dopaminergic neuron differentiation and function. J Neurochem 166:779–789
17. Cui LN, Dyball REJ (1996) Synaptic input from the retina to the suprachiasmatic nucleus changes with the light-dark cycle in the Syrian hamster. J Physiol 497: 483
18. Carlander B, Puech-Cathala AM, Jaussent I et al (2011) Low Vitamin D in narcolepsy with cataplexy. PLoS One 6:e20433
19. Snyder E, Cai B, DeMuro C, Morrison MF, Ball W (2018) A new single-item sleep quality scale: results of psychometric evaluation in patients with chronic primary insomnia and depression. J Clin Sleep Med 14: 1849–1857
20. Cappuccio FP, D'Elia L, Strazzullo P, Miller MA (2010) Sleep duration and all-cause mortality: a systematic review and meta-analysis of prospective studies. Sleep 33:585–592
21. Ikonte CJ, Mun JG, Reider CA, Grant RW, Mitmesser SH (2019) Micronutrient inadequacy in short sleep: analysis of the NHANES 2005–2016. Nutrients 11
22. Massa J, Stone KL, Wei EK et al (2015) Vitamin D and actigraphic sleep outcomes in older community-dwelling men: the MrOS sleep study. Sleep 38:251–257
23. Al-Shawwa B, Ehsan Z, Ingram DG (2020) Vitamin D and sleep in children. J Clin Sleep Med 16:1119–1123
24. Asuman C, Aydan K, Meric KN (2024) Role of vitamin D in the association between pre-stroke sleep quality and poststroke depression and anxiety. Sleep Breath 28:841–848
25. McCarty DE (2010) Resolution of hypersomnia following identification and treatment of Vitamin D deficiency. J Clin Sleep Med 6:605–608
26. McCarty DE, Reddy A, Keigley Q, Kim PY, Marino AA (2012) Vitamin D, race, and excessive daytime sleepiness. J Clin Sleep Med 8:693–697
27. Jung YS, Chae CH, Kim YO et al (2017) The relationship between serum vitamin D levels and sleep quality in fixed day indoor field workers in the electronics manufacturing industry in Korea. Ann Occup Environ Med 29:1–9
28. Mirzaei-Azandaryani Z, Abdolalipour S, Mirghafourvand M (2022) The effect of vitamin D on sleep quality: a systematic review and meta-analysis. Nutr Health 28:515–526
29. Huang W, Shah S, Long Q, Crankshaw AK, Tangpricha V (2013) Improvement of pain, sleep, and quality of life in chronic pain patients with vitamin D supplementation. Clin J Pain 29:341–347
30. Majid MS, Ahmad HS, Bizhan H, Hosein HZM, Mohammad A (2018) The effect of vitamin D supplement on the score and quality of sleep in 20–50 year-old people with sleep disorders compared with control group. Nutr Neurosci 21:511–519
31. McCarthy MS, Elshaw EB, Szekely BM, Raju D (2019) A prospective cohort study of Vitamin D supplementation in AD soldiers: preliminary findings. Mil Med 184:498–505
32. Ohayon MM (2002) Epidemiology of insomnia: what we know and what we still need to learn. Sleep Med Rev 6:97–111
33. Dempsey JA, Veasey SC, Morgan BJ, O'Donnell CP (2010) Pathophysiology of sleep apnea. Physiol Rev 90:47–112
34. Upala S, Sanguankeo A (2015) Association between 25-Hydroxyvitamin D and obstructive sleep apnea: a systematic review and meta-analysis. J Clin Sleep Med 11:1347
35. Goswami U, Ensrud KE, Paudel ML et al (2016) Vitamin D concentrations and obstructive sleep apnea in a multicenter cohort of older males. Ann Am Thorac Soc 13:712–718
36. Archontogeorgis K, Economou NT, Bargiotas P et al (2024) Sleepiness and Vitamin D levels in patients with obstructive sleep apnea. Healthcare (Basel, Switzerland) 12:698
37. Kerley CP, Hutchinson K, Bolger K, McGowan A, Faul J, Cormican L (2016) Serum Vitamin D is significantly inversely associated with disease severity in Caucasian adults with obstructive sleep apnea syndrome. Sleep 39:293–300

38. Archontogeorgis K, Nena E, Papanas N et al (2017) Vitamin D levels in middle-aged patients with obstructive sleep Apnoea syndrome. Curr Vasc Pharmacol 16: 289–297
39. Salepci B, Caglayan B, Nahid P et al (2017) Vitamin D deficiency in patients referred for evaluation of obstructive sleep apnea. J Clin Sleep Med 13:607–612
40. Compston JE, Vedi S, Ledger JE, Webb A, Gazet JC, Pilkington TR (1981) Vitamin D status and bone histomorphometry in gross obesity. Am J Clin Nutr 34:2359–2363
41. Patil SP, Schneider H, Schwartz AR, Smith PL (2007) Adult obstructive sleep apnea: pathophysiology and diagnosis. Chest 132:325–337
42. Liguori C, Romigi A, Izzi F et al (2015) Continuous positive airway pressure treatment increases serum Vitamin D levels in male patients with obstructive sleep apnea. J Clin Sleep Med 11:603–607
43. Siachpazidou DI, Kotsiou OS, Stavrou V et al (2021) Serum vitamin D levels in patients with obstructive sleep apnea syndrome and level changes after continuous positive airway pressure therapy. Sleep Breath 25:657–668
44. Theorell-Haglöw J, Hoyos CM, Phillips CL et al (2018) Changes of vitamin D levels and bone turnover markers after CPAP therapy: a randomized sham-controlled trial. J Sleep Res 27:e12606
45. Shahid A, Wilkinson K, Marcu S, Shapiro CM (2001) STOP, THAT and one hundred other sleep scales
46. Larsen AU, Hopstock LA, Jorde R, Grimnes G (2022) Associations of serum 25-hydroxyvitamin D and subjective sleep measures in an arctic population: insights from the population-based Tromsø Study. Sleep Med X 4:100056
47. Ribeiro JD, Pease JL, Gutierrez PM et al (2012) Sleep problems outperform depression and hopelessness as cross-sectional and longitudinal predictors of suicidal ideation and behavior in young adults in the military. J Affect Disord 136:743–750
48. Chesson A, Hartse K, Anderson WMD et al (2000) Practice parameters for the evaluation of chronic insomnia. An American Academy of Sleep Medicine report. Standards of Practice Committee of the American Academy of Sleep Medicine. Sleep 23:237–241
49. Jiang Y, Ge S, Wang C, Jin C, Zhao Y, Liu Q (2024) Causal relationship between micronutrient and sleep disorder: a Mendelian randomization study. Nat Sci Sleep 16:1267–1277
50. Bahrami A, Rezaeitalab F, Farahmand SK et al (2021) High-dose Vitamin D supplementation and improvement in cognitive abilities, insomnia, and daytime sleepiness in adolescent girls. Basic Clin Neurosci 12:339–348
51. Sharifan P, Khoshakhlagh M, Khorasanchi Z et al (2020) Efficacy of low-fat milk and yogurt fortified with encapsulated vitamin D3 on improvement in symptoms of insomnia and quality of life: evidence from the SUVINA trial. Food Sci Nutr 8:4484–4490
52. Larsen AU, Hopstock LA, Jorde R, Grimnes G (2021) No improvement of sleep from vitamin D supplementation: insights from a randomized controlled trial. Sleep Med X. 3:100040
53. Trenkwalder C, Paulus W (2010) Restless legs syndrome: pathophysiology, clinical presentation and management. Nat Rev Neurol 6:337–346
54. Ekbom K, Ulfberg J (2009) Restless legs syndrome. J Intern Med 266:419–431
55. Mansourian M, Rafie N, Khorvash F, Hadi A, Arab A (2020) Are serum vitamin D, calcium and phosphorous associated with restless leg syndrome? A systematic review and meta-analysis. Sleep Med 75:326–334
56. Wali S, Alsafadi S, Abaalkhail B et al (2018) The association between Vitamin D level and restless legs syndrome: a population-based case-control study. J Clin Sleep Med 14:557–564
57. Calvello R, Cianciulli A, Nicolardi G et al (2017) Vitamin D treatment attenuates Neuroinflammation and dopaminergic neurodegeneration in an animal model of Parkinson's disease, shifting M1 to M2 microglia responses. J Neuroimmune Pharmacol 12: 327–339
58. Turjanski N, Lees AJ, Brooks DJ (1999) Striatal dopaminergic function in restless legs syndrome: 18F-dopa and 11C-raclopride PET studies. Neurology 52:932–937
59. Kang J, Park M, Kim T (2024) Vitamin D reduces GABA-positive astrocytes in the 5xFAD mouse model of Alzheimer's disease. J Alzheimer's Dis 97:1939–1950
60. Wali S, Shukr A, Boudal A, Alsaiari A, Krayem A (2015) The effect of vitamin D supplements on the severity of restless legs syndrome. Sleep Breath 19: 579–583
61. Tutuncu M, Tutuncu M (2020) The effect of vitamin D on restless legs syndrome: prospective self-controlled case study. Sleep Breath 24:1101–1106
62. Burgess CR, Scammell TE (2012) Narcolepsy: neural mechanisms of sleepiness and cataplexy. J Neurosci 32:12305–12311
63. Ohayon M, Wickwire EM, Hirshkowitz M et al (2017) National Sleep Foundation's sleep quality recommendations: first report. Sleep Heal 3:6–19
64. Abboud M (2022) Vitamin D supplementation and sleep: a systematic review and meta-analysis of intervention studies. Nutrients:14
65. Kerley CP, Hutchinson K, Bramham J, McGowan A, Faul J, Cormican L (2017) Vitamin D improves selected metabolic parameters but not neuropsychological or quality of life indices in OSA: a pilot study. J Clin Sleep Med 13:19–26
66. Rosen CJ, Abrams SA, Aloia JF et al (2012) IOM Committee Members respond to Endocrine Society Vitamin D guideline. J Clin Endocrinol Metab 97: 1146–1152
67. Demay MB, Pittas AG, Bikle DD et al (2024) Vitamin D for the prevention of disease: an Endocrine Society clinical practice guideline. J Clin Endocrinol Metab 109(8):1907–1947. https://doi.org/10.1210/clinem/dgae290
68. Vasil N, Razzaque MS (2025) Decoding vitamin D intake: how much is too much? Adv Exp Med Biol. https://doi.org/10.1007/978-3-032-04357-3_2

Vitamin D and Fracture Healing

Md S. Zaman and M. I. Ullah

Abstract

Maintaining bone health involves complex interactions among nutrients, hormones, and physiological mechanisms, with vitamin D playing a pivotal role. Vitamin D is essential for the absorption of calcium and phosphate, which are critical minerals for bone development, maintenance, and repair. Disruptions in mineral ion homeostasis can lead to conditions such as hypocalcemia, osteoporosis, osteomalacia, and rickets, which are prevalent worldwide. Inadequate vitamin D impairs bone health, increasing the risk of fractures and catabolic bone diseases. This chapter elaborates on the synthesis, activation, and function of vitamin D, highlighting its essential role in bone health, from supporting growth in children to maintaining bone strength in adults. It also elaborates on how vitamin D enhances healing after a fracture by regulating the immune response, inflammation, and mineralization process.

M. S. Zaman (✉)
Department of Biological Sciences, Alcorn State University, Lorman, MS, USA

Department of Biology, South Texas College, McAllen, TX, USA
e-mail: zaman@alcorn.edu

M. I. Ullah
Division of Internal Medicine and Hypertension, University of Mississippi Medical Center, Jackson, MS, USA

Keywords

Vitamin D activation · Ergocalciferol · Cholecalciferol · Calcium regulation · Phosphate regulation · Bone growth · Fracture healing

Introduction

Maintaining bone health involves a complex interplay of nutrients, hormones, and bone cells. Vitamin D plays a pivotal role as a key regulator of calcium and phosphate metabolism, which are essential for bone development, maintenance, and repair [1]. Recent research has expanded our understanding of its role beyond just bone mineralization, highlighting its broader functions [2–4]. However, the primary focus remains on how vitamin D helps maintain optimal levels of calcium and phosphate, two critical minerals that ensure bone strength, growth, and repair, to support the structure and function of the skeletal system. Low vitamin D levels can impair calcium and phosphate homeostasis, leading to hypocalcemia, osteoporosis, osteomalacia, and rickets. These disorders affect millions of people worldwide, emphasizing the importance of maintaining adequate vitamin D levels [5, 6].

Calcium supports bone health and is crucial for muscle contraction, nerve impulse conduction, and blood clotting [7–9]. It is essential for the development, growth, and maintenance of strong

M. S. Razzaque (ed.), *Vitamin D Function*, Advances in Experimental Medicine and Biology 1493,
https://doi.org/10.1007/978-3-032-04357-3_8

bones and teeth. Approximately 99% of the body's calcium is stored in bones and teeth, providing structural support and hardness. It is involved in various cellular processes, including enzyme activation, hormone secretion, and cell signaling, which are critical for the proper function of the body. It also helps maintain the body's acid-base balance by acting as a buffer, keeping the body's pH levels within a narrow, healthy range. Phosphate, although less frequently discussed, combines calcium to form *hydroxyapatite crystals*, which are essential for bone hardness and structural integrity. It is a vital component of ATP (adenosine triphosphate), which is the energy currency of cells. Some phospholipid molecules make up cell membranes, providing structure and protection to cells. Imbalances in phosphate regulation, often associated with insufficient vitamin D, can lead to muscle weakness, poor nerve conduction, respiratory failure, irregular heartbeats, as well as bone deformities and increased susceptibility to fractures [10, 11].

This chapter describes the role of vitamin D in maintaining calcium and phosphate homeostasis and its impact on bone health, from promoting bone growth in children to enhancing fracture healing and managing osteoporosis in older adults.

Vitamin D Structure and Synthesis

The Structure of Vitamin D

Vitamin D is a group of fat-soluble steroids. It has two main forms: vitamin D2 (ergocalciferol) and vitamin D3 (cholecalciferol). Both forms share a similar structure, with slight differences in their side chains that distinguish D2 from D3. It has a secosteroid backbone, which means that it is a steroid with one of its rings broken. The structure is derived from cholesterol and shares similarities with that of steroid hormones [12] (Fig. 1).

Endogenous Synthesis of Vitamin D

Vitamin D synthesis in the body occurs through the skin upon exposure to ultraviolet B (UVB) radiation from sunlight. It begins when the skin is exposed to ultraviolet B (UVB) rays from sunlight which converts the *7-dehydrocholesterol* present in the skin into vitamin D3 (*cholecalciferol*). Vitamin D may also enter the body via dietary sources in the form of vitamin D2 (*ergocalciferol*) or D3. Ergocalciferol in food is almost identical to the cholecalciferol formed in the skin, except for the substitution of one or more atoms on the side chain that do not affect its function (Fig. 1). It then undergoes further transformation in the liver and kidneys to become the biologically active form of vitamin D, 1,25-dihydroxyvitamin D [1,25$(OH)_2$D], commonly known as calcitriol [13–15] (Fig. 2). Calcitriol is crucial for regulating calcium and phosphate absorption in the intestines, maintaining healthy levels of these minerals in the blood, and supporting bone remodeling to ensure continuous bone renewal and strength [16]. It binds to vitamin D receptors (VDRs) in the tissues of bones, kidneys, and intestines. These receptors increase calcium and phosphate absorption from food, promote calcium reabsorption in the kidneys, and regulate the release of calcium from bones [17].

Factors Influencing Endogenous Vitamin D Synthesis

Geographical location Individuals near the equator receive better year-round UVB exposure than those at northern latitudes do, especially during the winter months. This could affect vitamin D production [18].

Time of day UVB rays are most intense between late morning and midafternoon, making this period optimal for vitamin D production [19].

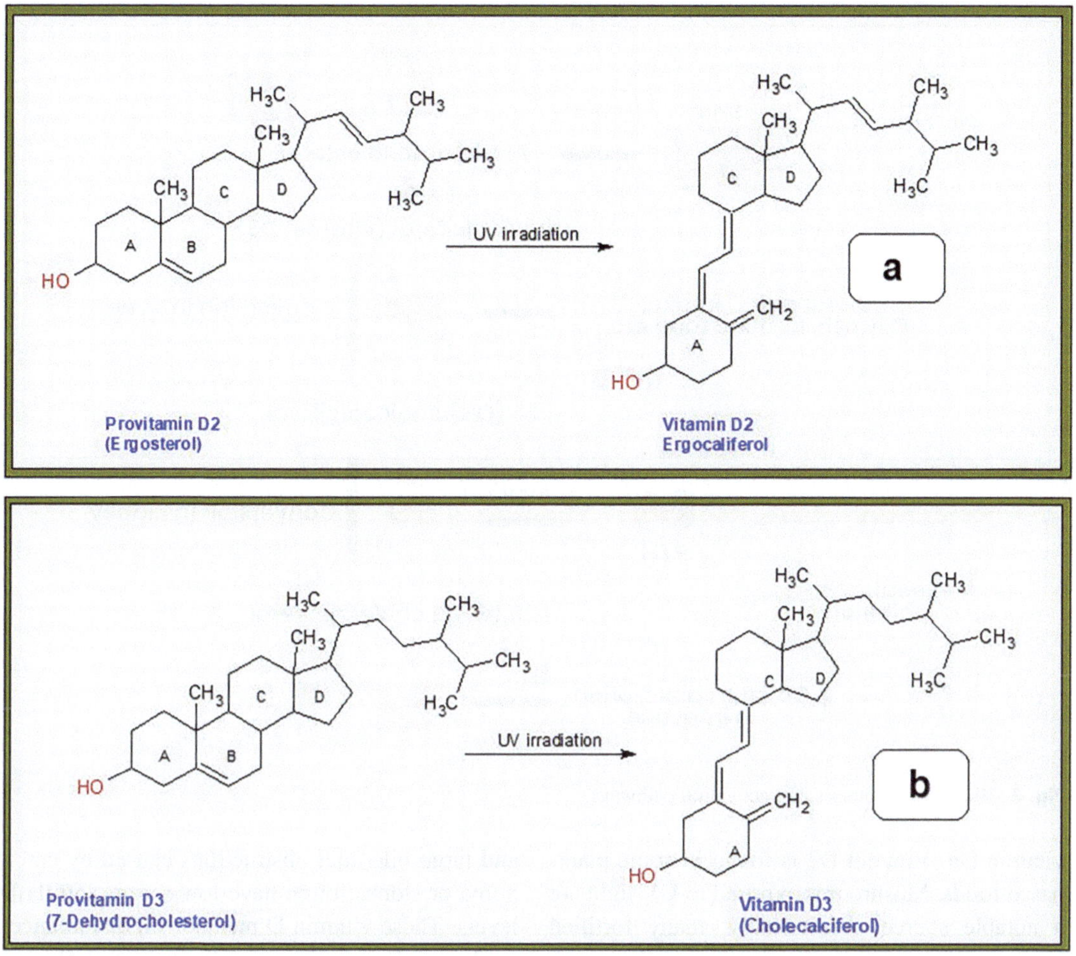

Fig. 1 UV radiation exposure converts (**a**) provitamin D2 (ergosterol) into vitamin D2 (ergocalciferol: $C_{28}H_{44}O$) and (**b**) provitamin D3 (7-dehydrocholesterol) into D3 (cholecalciferol: $C_{27}H_{44}O$). Both forms share similar chemical structures with minor alterations in their side chains

Skin tone Melanin acts as a natural barrier to UVB rays. Therefore, individuals with darker skin require more sun exposure to produce adequate vitamin D than those with lighter skin [20].

Age Aging reduces the ability of the skin to synthesize vitamin D efficiently, making it challenging for older adults to maintain sufficient levels of vitamin D [21].

Sunscreen and clothing Sunscreening with an SPF (sun protection factor) of 30 or higher can reduce vitamin D synthesis by 95% or more. Clothing can also block UVB rays [22].

Despite these variables, sunlight remains the most effective and natural method for maintaining vitamin D levels. For most people, 10–30 min of sun exposure to the face, arms, or legs several times a week is usually sufficient. However, when sunlight is inadequate or when exposure is limited, dietary sources and supplements are recommended to ensure sufficient vitamin D levels [23]. Vitamin D3 is the most potent form and is found primarily in animal-based foods [24]. Major sources include fatty fish such as salmon, mackerel, and sardines. Cod liver oil is a highly concentrated source of vitamin D3. Egg yolks and beef liver are also good sources of

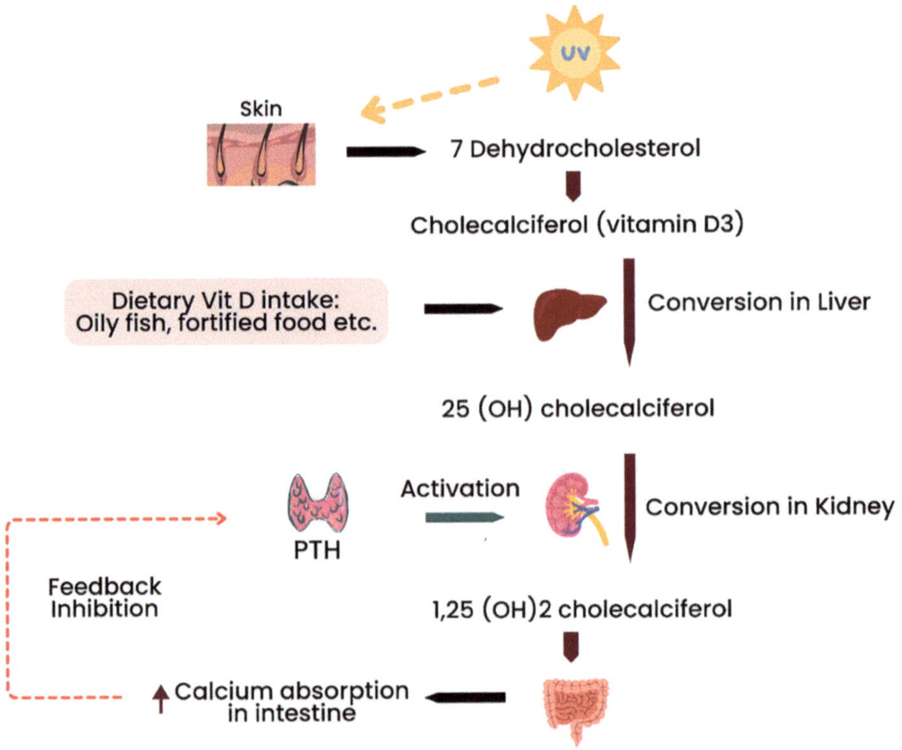

Fig. 2 Vitamin D synthesis and activation pathways

vitamin D3. Vitamin D2 is found in some plant-based foods. Mushrooms exposed to UV light are a notable source. Additionally, many fortified foods, such as milk, cereals, plant-based milk alternatives (such as almond or soy milk), and orange juice, contain added vitamin D2. Although it is less potent than vitamin D3, vitamin D2 is still beneficial, especially for individuals following vegetarian or vegan diets [12].

Vitamin D Activation in Chronic Disease

The two-step vitamin D activation process in the body is necessary for it to become biologically active, which may be impaired if the liver or kidneys are not functioning properly. In chronic liver disease, the production of 25(OH)D is negatively affected. Patients with various liver disorders, such as primary biliary cirrhosis, alcoholic liver disease, chronic active liver disease, and large bile-duct obstructions caused by carcinoma or stones, often have low serum 25(OH)D levels. These vitamin D metabolism disturbances in chronic liver and biliary diseases are linked to disrupted calcium and phosphate homeostasis, which clinically present as hepatic osteodystrophy. This condition includes osteomalacia, which can be complicated by secondary hyperparathyroidism, osteoporosis, and periosteal new bone formation [25].

In chronic kidney disease (CKD), there is decreased renal conversion of 25(OH)D (calcidiol) to 1,25$(OH)_2$D (calcitriol). This leads to lower levels of the active form of vitamin D, which is crucial for calcium and phosphate absorption to maintain bone health. In addition, decreased levels of active vitamin D lead to stimulation of the parathyroid glands to produce more parathyroid hormone (PTH) [26]. This secondary hyperparathyroidism leads to increased bone resorption and further complications in bone metabolism. Patients with CKD may also often have lower levels of vitamin D due to reduced dietary intake from poor appetite, as well as limited sun exposure from decreased mobility. This can further aggravate the risk of bone disease and other complications. CKD increases the levels of fibroblast growth factor 23 (FGF-23), which inhibits the conversion of vitamin D to its active form and can lead to further complications in mineral and bone metabolism [27]. Thus, CKD may ultimately cause osteomalacia (inadequate mineralization causing soft bone) in adults and rickets in children and, over time, contribute to osteoporosis which makes bones prone to fractures.

Regulation of Vitamin D Activation by PTH

PTH plays a central role in regulating the activation of vitamin D, particularly when the body detects low calcium levels in the blood. Once released, PTH stimulates bone resorption, causing the release of calcium into the extracellular fluid. It also increases the reabsorption of calcium by the renal tubules, leading to decreased excretion of calcium [13]. By promoting vitamin D activation, PTH ensures that the body has sufficient active vitamin D to absorb and utilize calcium efficiently, especially during periods of low dietary calcium intake or increased physiological demands such as growth, pregnancy, or aging [9, 18]. The regulation of vitamin D activation is part of a feedback system that maintains calcium and phosphate balance in the body (Fig. 3) [13, 28]. In hypocalcemia, the parathyroid glands release more PTH to promote the activation of vitamin D in the kidneys, signals bones to release stored calcium, and reduces calcium loss in the urine. Hypocalcemia also causes increased production of calcitriol, which, in turn, increases calcium absorption from the intestines to help increase blood calcium levels. These feedback mechanisms help normalize calcium level. As the calcium level increases, a negative feedback loop is triggered. High calcium concentrations signal that the parathyroid glands reduce PTH secretion. With less PTH, vitamin D activation slows down, limiting calcitriol production. This feedback prevents hypercalcemia, which could lead to complications such as kidney stones, soft tissue, or cardiovascular calcification.

Phosphate levels also influence vitamin D activation. Elevated phosphate can inhibit calcitriol production to balance calcium and phosphate levels. High phosphate reduces 1α-hydroxylase activity in the kidneys to reduce vitamin D activation. The kidneys also excrete more phosphate under the influence of PTH. This prevents excess phosphate buildup in the blood, ensuring calcium and phosphate equilibrium, essential for adequate bone mineralization.

Vitamin D and Bone Growth

Vitamin D is essential for bone development, particularly during childhood and adolescence, when the skeletal system undergoes rapid growth. As discussed earlier, vitamin D supports bone growth by regulating the absorption and balance of calcium and phosphate, which are critical for bone mineralization. This process occurs primarily at the epiphyseal plates (growth plates), where new bone cells are produced and mineralized to ensure proper bone lengthening and strength. At the cellular level, vitamin D enhances the intestinal absorption of calcium and phosphate, ensuring that enough of these minerals are available in the bloodstream to be incorporated into the bone matrix. In the absence of adequate vitamin D, calcium absorption is significantly reduced, leading to lower calcium levels. This stimulates the release of PTH. Elevated PTH increases bone resorption—the breakdown of bone tissue to release calcium into the blood, ultimately compromising bone formation and strength [9, 29]. Additionally, vitamin D promotes the differentiation of osteoblasts, the cells responsible for the formation of new bones. This function is particularly important during the rapid skeletal growth that occurs in childhood and adolescence. In the absence of sufficient vitamin D, bone mineralization is impaired, leading to weak and poorly structured bones, as commonly encountered in rickets and osteomalasia [30–33].

For many individuals, obtaining sufficient vitamin D through sunlight exposure or diet alone can be challenging; thus,

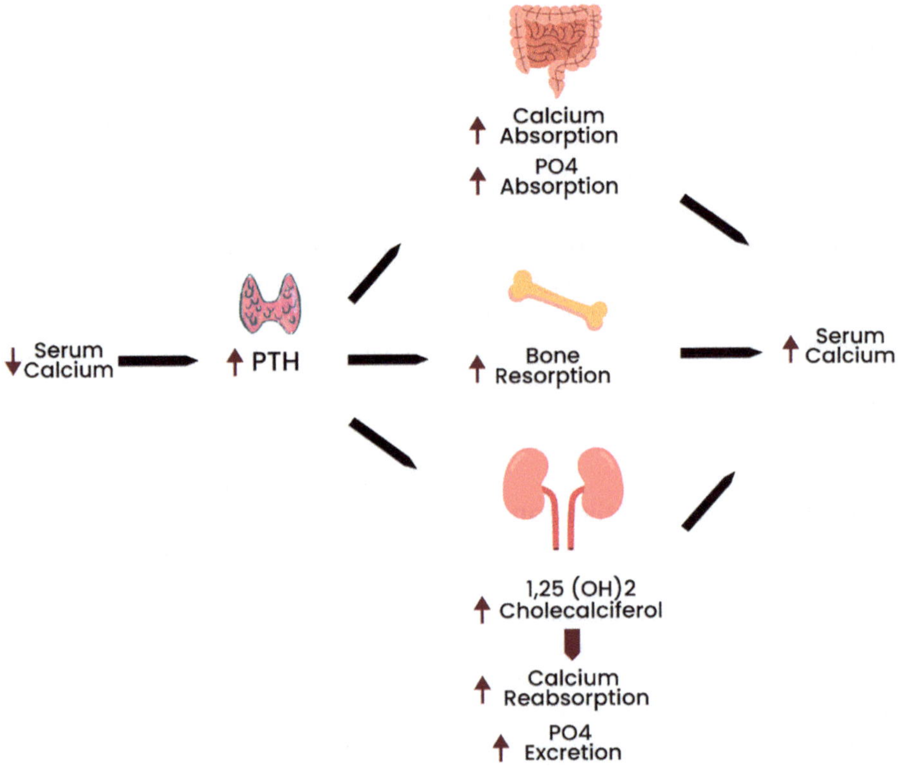

Fig. 3 Feedback mechanism to control calcium and phosphate balance

supplementation often becomes necessary. For example, infants, who are exclusively breastfed, may require supplemental vitamin D since breast milk contains only small amounts of vitamins. The American Academy of Pediatrics recommends that breastfed and partially breastfed infants receive 400 international units (IU) of vitamin D daily, starting within the first few days of life [34]. Additionally, children living in northern latitudes, where sunlight is less intense, may benefit from vitamin D supplementation, especially during the winter months when UVB radiation is insufficient for skin synthesis [35].

Vitamin D in Fracture Healing

Fracture healing is a complex process that involves multiple phases aimed at restoring the structural and functional integrity of the bone. Vitamin D plays a critical role in the bone repair process by enhancing the bioavailability of calcium and phosphate, the two critical minerals required for bone repair. The active form of vitamin D, $1,25(OH)_2D$, facilitates calcium and phosphate absorption in the intestines and maintains their levels in the bloodstream [36]. Vitamin D promotes osteoblast function, such as bone matrix synthesis and mineralization, by supporting their differentiation and activities [37]. Moreover, it helps regulate the balance between osteoblast and osteoclast activity, promoting bone formation while limiting excessive bone resorption [38]. Research has consistently shown that patients with higher vitamin D levels experience faster recovery times, better fracture healing, and lower rates of complications than those with vitamin D deficiency [39, 40].

Phases of Fracture Healing

The healing of a bone fracture occurs in three distinct but overlapping stages: (1) the inflammatory phase, (2) the reparative phase, and (3) the remodeling phase. Each phase relies on the precise coordination of cellular and molecular processes, in which vitamin D plays an integral part [41, 42].

Inflammatory phase The inflammatory phase begins immediately after the fracture occurs and lasts for several days. This phase is characterized by the formation of a blood clot (hematoma) at the fracture site, which serves as a platform for recruiting inflammatory cells, including macrophages, neutrophils, and fibroblasts. These cells release cytokines and growth factors that promote the removal of damaged tissue and initiate the healing process. The inflammatory phase primarily focuses on preparing the fracture site for repair and setting the stage for subsequent bone formation. Vitamin D supports this phase by regulating the immune response and reducing excessive inflammation, which can delay healing if not properly controlled.

Reparative phase The reparative phase begins after the inflammatory phase and may last from a few days to several weeks. During this phase, a soft callus made of collagen and cartilage is formed around the fracture site and gradually transforms into a hard, mineralized bone callus. Osteoblast cells promote the deposition of new bone matrix and initiate mineralization with calcium and phosphate. Vitamin D actively plays a role in this process by promoting the absorption of calcium and phosphate from the intestines and regulating their incorporation into the bone matrix. This ensures that sufficient minerals are available to form healthy bones.

Remodeling phase The final phase of fracture healing is the remodeling phase, which can last for months or even years. The newly formed bone is reshaped and strengthened during this phase to restore its native structure and mechanical properties. Osteoclasts resorb excess or poorly aligned bone tissue, while osteoblasts continue to deposit new bone tissue in a controlled manner. Vitamin D supports this remodeling phase by maintaining calcium and phosphate homeostasis. Without adequate vitamin D, the remodeling process may be compromised, leading to delayed healing or poor bone quality.

Impact of Vitamin D Sufficiency on Fracture Recovery

Faster recovery and improved healing Patients with adequate vitamin D levels [serum 25 (OH)D levels above 20 ng/mL] tend to heal more quickly after fracture [40]. Studies indicate that patients with vitamin D sufficiency have shorter recovery times and a greater likelihood of complete fracture healing within the expected timeframe, than do those with vitamin D deficiency [43, 44]. These findings suggest that vitamin D promotes efficient bone mineralization and callus formation during the reparative phase.

Reduction in nonunion and delayed healing Vitamin D deficiency is associated with an increased risk of delayed fracture healing and nonunion, a condition in which the fracture fails to heal properly. Patients with lower vitamin D levels are more likely to experience delayed union or nonunion than are those with higher vitamin D levels. Thus, maintaining optimal vitamin D levels is critical for reducing the risk of these complications [43].

Fracture risk reduction and postfracture outcomes Individuals with adequate vitamin D intake have a significantly lower risk of fractures than do those with insufficient vitamin D [45]. Thus, optimum vitamin D levels may help prevent fractures and promote effective healing after injury.

Vitamin D Supplementation

Vitamin D supplementation is a well-established strategy for the prevention and management of bone health, especially in individuals at risk for vitamin D deficiency, such as older adults, postmenopausal women, and individuals with limited sun exposure. Supplementing vitamin D helps ensure adequate calcium absorption, reduce bone resorption, and maintain bone density, thereby reducing the risk of fractures and osteoporosis [10, 46].

Potential Benefits of Vitamin D Supplementation

Studies have shown that vitamin D supplementation, often combined with calcium, can significantly reduce the risk of fractures in osteoporosis patients [32]. Vitamin D supplementation helps maintain optimal serum levels of calcidiol [25 (OH)D], the marker used to assess vitamin D status. This is crucial for preventing secondary hyperparathyroidism which accelerates bone loss [47]. By improving calcium absorption and reducing PTH levels, vitamin D supplementation increases bone strength and reduces the likelihood of falls and fractures, particularly in elderly populations [48]. The *Women's Health Initiative* trial, which investigated the effects of vitamin D and calcium supplementation in postmenopausal women, revealed that those who received 1000 mg of calcium and 400 IU of vitamin D daily had a modest but significant reduction in hip fracture risk compared with those who received a placebo [49]. Another meta-analysis showed that vitamin D supplementation, particularly when combined with calcium, significantly reduced the risk of hip fractures and total fractures in older adults. The benefits were most pronounced in individuals with vitamin D deficiency [50]. In addition to its role in reducing fracture risk, vitamin D has been shown to improve muscle strength and balance, which are critical for preventing falls and are major causes of fractures in elderly individuals with osteoporosis. Vitamin D supplementation can reduce the risk of falls by improving neuromuscular function, thereby indirectly reducing the risk of fractures. This dual benefit of vitamin D, which supports bone strength and reduces falls, makes it an essential component of osteoporosis care [51–53].

Dosage Recommendations

The recommended daily intake of vitamin D varies depending on age, health status, and individual risk factors for deficiency. Supplements are available in both vitamin D2 and D3 forms, with vitamin D3 being the preferred option owing to its superior potency and longer half-life. Compared with vitamin D2, vitamin D3 is approximately 87% more effective at increasing and maintaining serum 25(OH)D levels and leads to two to three times greater vitamin D storage [24]. The serum 25(OH)D level at or above 20 ng/mL is generally considered sufficient to support calcium absorption and bone health [46]. However, some argue that higher serum 25 (OH)D levels, in the range of 30–50 ng/mL, may be optimal for promoting bone growth and mineralization, especially in children at risk of vitamin D deficiency. These higher levels may be particularly important in populations with limited sun exposure, darker skin pigmentation (which reduces the ability of the skin to synthesize vitamin D), or underlying health conditions that affect vitamin D metabolism [54, 55].

For most adults, the recommended dietary allowance (RDA) for vitamin D is 600–800 international units (IU) per day, with higher doses recommended for older adults. The *Endocrine Society* suggests that individuals at risk for vitamin D deficiency, such as older adults and those with osteoporosis, may benefit from higher doses of 1000–2000 IU per day to maintain optimal serum 25(OH)D levels (above 30 ng/mL), which is considered sufficient for bone health. The usual treatment for severe vitamin D deficiency is to take 50,000 IU of vitamin D once a week for 6–8 weeks or until the 25(OH)D level exceeds 30 ng/mL. Patients typically take 1000–2000 IU daily to maintain their levels [56–58].

Potential Risks of Vitamin D Supplementation

While vitamin D supplementation is generally safe, excessive doses can lead to hypercalcemia (elevated calcium levels in the blood), which may cause kidney stones, cardiovascular issues, and other health problems. Therefore, excessive supplementation should be avoided, and vitamin D and calcium levels should be monitored, especially in patients with kidney disease or other conditions affecting calcium metabolism. In general, patients should avoid taking more than 1500 mg of calcium per day [59]. The Endocrine Society recommends that doses of vitamin D should not exceed 10,000 IU per day for adults aged 19 years and older [60].

Conclusions

Vitamin D promotes bone mineralization, which is the largest storage of calcium and phosphate in the body. Vitamin D promotes bone growth, healing, and remodeling by maintaining a balance between the activities of osteoblasts and osteoclasts. Without sufficient vitamin D, bones can become thin, brittle, or malformed, leading to conditions such as rickets in children and osteomalacia and osteoporosis in adults. Thus, optimum vitamin D status should be maintained to prevent fractures and promote effective bone healing after injury through supplementations. For newer vitamin D guidelines for testing and dosing, readers are recommended to a recent publication [57] and chapter "Decoding Vitamin D Intake: How Much Is Too Much?" of this book [61].

References

1. Girgis CM, Clifton-Bligh RJ, Hamrick MW, Holick MF, Gunton JE (2013) The roles of vitamin D in skeletal muscle: form, function, and metabolism. Endocr Rev 34(1):33–83. https://doi.org/10.1210/ER.2012-1012
2. Charoenngam N, Holick MF (2020) Immunologic effects of vitamin D on human health and disease. Nutrients 12(7):1–28. https://doi.org/10.3390/NU12072097
3. Mattioli AV, Coppi F, Severino P et al (2024) A personalized approach to vitamin D supplementation in cardiovascular health beyond the bone: an expert consensus by the Italian National Institute for Cardiovascular Research. Nutrients 17(1):115. https://doi.org/10.3390/NU17010115
4. Ullah MI, Uwaifo GI, Nicholas WC, Koch CA (2010) Does vitamin D deficiency cause hypertension? Current evidence from clinical studies and potential mechanisms. Int J Endocrinol 2010:1–11
5. Šromová V, Sobola D, Kaspar P (2023) A brief review of bone cell function and importance. Cells 12(21): 2576. https://doi.org/10.3390/CELLS12212576
6. Vimalraj S (2020) Alkaline phosphatase: structure, expression and its function in bone mineralization. Gene 754:144855. https://doi.org/10.1016/J.GENE.2020.144855
7. Vannucci L, Fossi C, Quattrini S et al (2018) Calcium intake in bone health: a focus on calcium-rich mineral waters. Nutrients 10(12):1930. https://doi.org/10.3390/NU10121930
8. Bootman MD, Collins TJ, Peppiatt CM et al (2001) Calcium signalling—an overview. Semin Cell Dev Biol 12(1):3–10. https://doi.org/10.1006/SCDB.2000.0211
9. Khundmiri SJ, Murray RD, Lederer E (2016) PTH and vitamin D. Compr Physiol 6(2):561–601. https://doi.org/10.1002/CPHY.C140071
10. Penido MGMG, Alon US (2012) Phosphate homeostasis and its role in bone health. Pediatr Nephrol 27(11):2039–2048. https://doi.org/10.1007/S00467-012-2175-Z
11. Subramanian R, Khardori R (2000) Severe hypophosphatemia. Pathophysiologic implications, clinical presentations, and treatment. Medicine 79(1): 1–8
12. Martineau AR, Thummel KE, Wang Z et al (2019) Differential effects of oral boluses of vitamin D2 vs vitamin D3 on vitamin D metabolism: a randomized controlled trial. J Clin Endocrinol Metab 104(12): 5831. https://doi.org/10.1210/JC.2019-00207
13. Hall JE, Hall ME (2021) Parathyroid hormone, calcitonin, calcium and phosphate metabolism, vitamin D, bone, and teeth. In: Guyton and hall textbook of medical physiology, 14th edn. Elsevier, pp 997–999
14. Ramasamy I, Vitamin D (2020) Metabolism and guidelines for vitamin D supplementation. Clin Biochem Rev 41(3):103–126. https://doi.org/10.33176/AACB-20-00006
15. Bland R, Walker EA, Hughes SV, Stewart PM, Hewison M (1999) Constitutive expression of 25-hydroxyvitamin D3-1alpha-hydroxylase in a transformed human proximal tubule cell line: evidence for direct regulation of vitamin D metabolism by calcium. Endocrinology 140(5):2027–2034. https://doi.org/10.1210/ENDO.140.5.6683
16. Bhattarai HK, Shrestha S, Rokka K, Shakya R (2020) Vitamin D, calcium, parathyroid hormone, and sex steroids in bone health and effects of aging. J Osteoporos 2020:9324505. https://doi.org/10.1155/2020/9324505

17. Irsik DL, Bollag WB, Isales CM (2021) Renal contributions to age-related changes in mineral metabolism. JBMR Plus 5(10):e10517. https://doi.org/10.1002/JBM4.10517
18. Wacker M, Holick MF (2013) Sunlight and vitamin D: a global perspective for health. Dermatoendocrinol 5(1):51–108. https://doi.org/10.4161/DERM.24494
19. Chung M, Balk EM, Brendel M et al (2009) Vitamin D and calcium: a systematic review of health outcomes. Evid Rep Technol Assess (Full Rep) (183):1–420. http://www.ncbi.nlm.nih.gov/pubmed/20629479. Accessed 17 June 2013
20. Webb AR, Kazantzidis A, Kift RC, Farrar MD, Wilkinson J, Rhodes LE (2018) Colour counts: sunlight and skin type as drivers of vitamin D deficiency at UK latitudes. Nutrients 10(4):457. https://doi.org/10.3390/NU10040457
21. Giustina A, Bouillon R, Dawson-Hughes B et al (2023) Vitamin D in the older population: a consensus statement. Endocrine 79(1):31–44. https://doi.org/10.1007/S12020-022-03208-3
22. Schalka S, dos Reis VMS, Cucé LC (2009) The influence of the amount of sunscreen applied and its sun protection factor (SPF): evaluation of two sunscreens including the same ingredients at different concentrations. Photodermatol Photoimmunol Photomed 25(4):175–180. https://doi.org/10.1111/J.1600-0781.2009.00408.X
23. Chang SW, Lee HC (2019) Vitamin D and health – the missing vitamin in humans. Pediatr Neonatol 60(3):237–244. https://doi.org/10.1016/J.PEDNEO.2019.04.007
24. Heaney RP, Recker RR, Grote J, Horst RL, Armas LAG (2011) Vitamin D(3) is more potent than vitamin D(2) in humans. J Clin Endocrinol Metab 96(3):E447–E452. https://doi.org/10.1210/JC.2010-2230
25. Wills MR, Savory J (1984) Vitamin D metabolism and chronic liver disease. Ann Clin Lab Sci 14(3):189–197. http://www.ncbi.nlm.nih.gov/pubmed/6329069. Accessed 12 Jan 2025.
26. Elmukhtar Habas S, Eledrisi M, Khan F, Elzouki ANY (2021) Secondary hyperparathyroidism in chronic kidney disease: pathophysiology and management. Cureus 13(7):e16388. https://doi.org/10.7759/CUREUS.16388
27. De Borst MH, Vervloet MG, Ter Wee PM, Navis G (2011) Cross talk between the renin-angiotensin-aldosterone system and vitamin D-FGF-23-klotho in chronic kidney disease. J Am Soc Nephrol 22(9):1603–1609. https://doi.org/10.1681/ASN.2010121251
28. Ritter CS, Slatopolsky E (2016) Phosphate toxicity in CKD: the killer among us. Clin J Am Soc Nephrol 11(6):1088. https://doi.org/10.2215/CJN.11901115
29. Wacker M, Holiack MF (2013) Vitamin D – effects on skeletal and extraskeletal health and the need for supplementation. Nutrients 5(1):111–148. https://doi.org/10.3390/NU5010111
30. van Driel M, van Leeuwen JPTM (2023) Vitamin D and bone: a story of endocrine and auto/paracrine action in osteoblasts. Nutrients 15(3):480. https://doi.org/10.3390/NU15030480
31. Sahay M, Sahay R (2012) Rickets-vitamin D deficiency and dependency. Indian J Endocrinol Metab 16(2):164. https://doi.org/10.4103/2230-8210.93732
32. Herdea A, Ionescu A, Dragomirescu MC, Ulici A (2023) Vitamin D—a risk factor for bone fractures in children: a population-based prospective case-control randomized cross-sectional study. Int J Environ Res Public Health 20(4):3300. https://doi.org/10.3390/IJERPH20043300
33. Cashman KD (2007) Vitamin D in childhood and adolescence. Postgrad Med J 83(978):230–235. https://doi.org/10.1136/PGMJ.2006.052787
34. Domenici R, Vierucci F (2022) Exclusive breastfeeding and vitamin D supplementation: a positive synergistic effect on prevention of childhood infections? Int J Environ Res Public Health 19(5):2973. https://doi.org/10.3390/IJERPH19052973
35. Huotari A, Herzig KH (2008) Vitamin D and living in northern latitudes—an endemic risk area for vitamin D deficiency. Int J Circumpolar Health 67(2–3):164–178. https://doi.org/10.3402/IJCH.V67I2-3.18258
36. Akimbekov NS, Digel I, Sherelkhan DK, Razzaque MS (2022) Vitamin D and phosphate interactions in health and disease. In: Phosphate metabolism. Springer, Cham, pp 37–46. https://doi.org/10.1007/978-3-030-91623-7_5
37. Sundar R, Bhagavandas Rai A, Naveen Kumar J, Devang Divakar D (2023) The role of vitamin D as an adjunct for bone regeneration: a systematic review of literature. Saudi Dent J 35(3):220–232. https://doi.org/10.1016/J.SDENTJ.2023.02.002
38. Choi IA, Umemoto A, Mizuno M, Park-Min KH (2024) Bone metabolism—an underappreciated player. NPJ Metab Health Dis 2(1):1–10. https://doi.org/10.1038/s44324-024-00010-9
39. Slobogean GP, Bzovsky S, O'Hara NN et al (2022) Effect of vitamin D3 supplementation on acute fracture healing: a phase II screening randomized double-blind controlled trial. JBMR Plus 7(1):e10705. https://doi.org/10.1002/JBM4.10705
40. Maier GS, Weissenberger M, Rudert M, Roth KE, Horas K (2021) The role of vitamin D and vitamin D deficiency in orthopaedics and traumatology—a narrative overview of the literature. Ann Transl Med 9(11):942. https://doi.org/10.21037/ATM-21-779
41. Wilkins KE (2005) Principles of fracture remodeling in children. Injury 36(Suppl 1):A3–A11. https://doi.org/10.1016/J.INJURY.2004.12.007
42. Marsell R, Einhorn TA (2011) The biology of fracture healing. Injury 42(6):551–555. https://doi.org/10.1016/J.INJURY.2011.03.031
43. Gorter EA, Krijnen P, Schipper IB (2017) Vitamin D status and adult fracture healing. J Clin Orthop Trauma 8(1):34–37. https://doi.org/10.1016/J.JCOT.2016.09.003
44. Chevalley T, Brandi ML, Cavalier E et al (2021) How can the orthopedic surgeon ensure optimal vitamin D status in patients operated for an osteoporotic fracture? Osteoporos Int 32(10):1921–1935. https://doi.org/10.1007/S00198-021-05957-9

45. Yao P, Bennett D, Mafham M et al (2019) Vitamin D and calcium for the prevention of fracture: a systematic review and meta-analysis. JAMA Netw Open 2(12): E1917789. https://doi.org/10.1001/JAMANETWORKOPEN.2019.17789
46. Giustina A, Bilezikian JP, Adler RA et al (2024) Consensus statement on vitamin D status assessment and supplementation: whys, whens, and hows. Endocr Rev 45(5):625–654. https://doi.org/10.1210/ENDREV/BNAE009
47. Burt LA, Billington EO, Rose MS, Raymond DA, Hanley DA, Boyd SK (2019) Effect of high-dose vitamin D supplementation on volumetric bone density and bone strength: a randomized clinical trial. JAMA 322(8):736–745. https://doi.org/10.1001/JAMA.2019.11889
48. Khazai N, Judd SE, Tangpricha V (2008) Calcium and vitamin D: skeletal and extraskeletal health. Curr Rheumatol Rep 10(2):110–117. https://doi.org/10.1007/S11926-008-0020-Y
49. Cauley JA, Chlebowski RT, Wactawski-Wende J et al (2013) Calcium plus vitamin D supplementation and health outcomes five years after active intervention ended: the Women's Health Initiative. J Womens Health (Larchmt) 22(11):915–929. https://doi.org/10.1089/JWH.2013.4270
50. Waterhouse M, Ebeling PR, McLeod DSA et al (2023) The effect of monthly vitamin D supplementation on fractures: a tertiary outcome from the population-based, double-blind, randomised, placebo-controlled D-Health trial. Lancet Diabetes Endocrinol 11(5): 324–332. https://doi.org/10.1016/S2213-8587(23)00063-3
51. Mendes MM, Botelho PB, Ribeiro H (2022) Vitamin D and musculoskeletal health: outstanding aspects to be considered in the light of current evidence. Endocr Connect 11(10):e210596. https://doi.org/10.1530/EC-21-0596
52. Kong SH, Jang HN, Kim JH, Kim SW, Shin CS (2022) Effect of vitamin D supplementation on risk of fractures and falls according to dosage and interval: a meta-analysis. Endocrinol Metab (Seoul) 37(2): 344–358. https://doi.org/10.3803/ENM.2021.1374
53. Ling Y, Xu F, Xia X et al (2021) Vitamin D supplementation reduces the risk of fall in the vitamin D deficient elderly: an updated meta-analysis. Clin Nutr 40(11):5531–5537. https://doi.org/10.1016/J.CLNU.2021.09.031
54. Bischoff-Ferrari HA, Giovannucci E, Willett WC, Dietrich T, Dawson-Hughes B (2006) Estimation of optimal serum concentrations of 25-hydroxyvitamin D for multiple health outcomes. Am J Clin Nutr 84(1): 18–28. https://doi.org/10.1093/AJCN/84.1.18
55. Melamed ML, Michos ED, Post W, Astor B (2008) 25-Hydroxyvitamin D levels and the risk of mortality in the general population. Arch Intern Med 168(15): 1629–1637. https://doi.org/10.1001/ARCHINTE.168.15.1629
56. Płudowski P, Kos-Kudła B, Walczak M et al (2023) Guidelines for preventing and treating vitamin D deficiency: a 2023 update in Poland. Nutrients 15(3):695. https://doi.org/10.3390/NU15030695
57. Demay MB, Pittas AG, Bikle DD et al (2024) Vitamin D for the prevention of disease: an Endocrine Society clinical practice guideline. J Clin Endocrinol Metab 109(8):1907–1947. https://doi.org/10.1210/clinem/dgae290
58. NIH. Vitamin D—health professional fact sheet. National Institute of Health (NIH). https://ods.od.nih.gov/factsheets/VitaminD-HealthProfessional/. Accessed 17 Jan 2025
59. Curhan GC, Willett WC, Rimm EB, Stampfer MJ (1993) A prospective study of dietary calcium and other nutrients and the risk of symptomatic kidney stones. N Engl J Med 328(12):833–838. https://doi.org/10.1056/NEJM199303253281203
60. Holick MF, Binkley NC, Bischoff-Ferrari HA et al (2011) Evaluation, treatment, and prevention of vitamin D deficiency: an Endocrine Society clinical practice guideline. J Clin Endocrinol Metab 96(7): 1911–1930. https://doi.org/10.1210/JC.2011-0385
61. Vasil N, Razzaque MS (2025) Decoding vitamin D intake: how much is too much? In: Razzaque MS (ed) Vitamin D function. Advances in experimental medicine and biology, vol 1493. Springer, Cham, pp 5–11

Vitamin D and Oral Cancer

Chrispinus Hakimu Mumena, Nuraly S. Akimbekov, Rajabu Sasi, Majuto Mlawa, and Majambo Hussein Mudhihiri

Abstract

Oral cancer has high mortality and recurrence rates. This is attributed to resistance to the process of regulated cell death and resistance to radiotherapy and chemotherapy. Oral squamous cell carcinoma (OSCC) accounts for the majority of oral cancers. Its prognosis varies with the site, clinical stage, and histological grade of the tumor. Owing to the variation in the biological behavior of OSCC, the existing prognostic parameters have some limitations in terms of providing precise information on prognosis. The remarkable biological effects of vitamin D against oral cancer have been documented in various studies. These include antiproliferative and anti-invasive effects, inhibition of angiogenesis, stimulation of mutual adherence of cells, and induction of apoptosis in OSCC. Furthermore, vitamin D increases the sensitivity of OSCC tumor cells to chemotherapy. This phenomenon has been observed in tumors that are resistant to chemotherapy. These findings indicate that the biological effects of vitamin D may provide significant prognostic value in oral cancer patients. In addition, the administration of vitamin D improves the quality of life of patients with inoperable oral cancer lesions in addition to the reported increase in tumor sensitivity to chemotherapy. A low level of vitamin D is associated with increased oral cancer risk. A high frequency of vitamin D deficiency and pleomorphism of the VDR have been reported in oral cancer patients. Poorly differentiated and aggressive OSCCs exhibit more CYP24A1 and VDR polymorphisms and less CYP27B1. Determination of the serum level of vitamin D and identification of VDR, CYP24A1, and CYP27B1 in oral cancer cells may be of significant prognostic value in oral cancer patients and may significantly improve the management of head and neck cancers.

Keywords

Humans · Quality of life · Vitamin D · Prognosis · Complement factor D · Vitamin D

C. H. Mumena (✉)
The Copperbelt University, School of Medicine, Department of Dental Clinical Sciences, Ndola, Zambia

The Gothenburg University, Sahlgrenska Academy, Institute of Clinical Sciences, Oncology, Gothenburg, Sweden
e-mail: Chrispinus.mumena@cbu.ac.zm

N. S. Akimbekov
International Center for Islamic Science and Innovation, Al-Farabi Kazakh National University, Almaty, Kazakhstan

Scientific-Practical Center, West Kazakhstan Marat Ospanov Medical University, Aktobe, Kazakhstan

Ecology Research Institute, Khoja Akhmet Yassawi International Kazakh-Turkish University, Turkistan, Kazakhstan

R. Sasi · M. Mlawa · M. H. Mudhihiri
The Copperbelt University, School of Medicine, Department of Dental Clinical Sciences, Ndola, Zambia

M. S. Razzaque (ed.), *Vitamin D Function*, Advances in Experimental Medicine and Biology 1493,
https://doi.org/10.1007/978-3-032-04357-3_9

deficiency · Mouth neoplasms · Apoptosis · Regulated cell death

Introduction

Oral cancer is the sixth most common cancer, accounting for 890,000 new cases and 450,000 deaths worldwide [1, 2]. Most neoplasms in the head and neck region are oral squamous cell carcinomas (OSCCs) [3–5]. The biological behavior of these lesions varies widely across different oral sites, posing a challenge to the consistency of the clinicopathological prognostic parameters used [3, 6]. Furthermore, resistance to apoptosis and radio- and chemotherapy is among the factors that promote tumor recurrence [7]. Despite improvements in surgery, immunotherapy, and radio- and chemotherapy, the prognosis of patients with HNC remains poor, as indicated by the 5-year survival ranging from only 40% to 50% [8]. Factors such as recurrence, resistance to apoptosis, and chemo-/radiotherapy, as well as late reporting whereby patients seek hospital care at an advanced stage when the curative option is limited, contribute to the short survival of patients [3–5, 7]. Clinical stage (TNM), lymph node metastasis, and histopathological grade are long-term and cardinal prognostic parameters used in decision-making for the treatment of oral tumors [5]. These parameters do not consistently predict the prognosis or inform the treatment outcomes of advanced and aggressive tumors precisely [5]. The lack of consistency calls for the exploration of more reliable prognostic parameters that reflect the biological behavior of advanced and aggressive oral cancers. This may strengthen the management of oral malignant tumors.

The classic risk factors for oral cancer are alcohol and tobacco use [4, 5, 9, 10]. With trade globalization, movements and cultural infusion have improved worldwide, leading to increased availability and consumption of cigarettes and alcohol, increasing the incidence of oral cancer in developing countries [4, 5, 9, 10]. The two combined factors account for 80% of oral cancer risk [9]. Other risk factors include HPV infection, betel quid chewing [4–6], and dietary deficiencies, such as folate and vitamins A, C, D, and E [5, 6, 9]. Accumulating evidence indicates that low vitamin D is associated with oral cancer [9, 11–13]. The studied patients with oral cancer present with VDR gene polymorphisms, mutations, or deletions [9, 13]. The role of vitamin D in anticancer effects is supported by evidence of single nucleotide polymorphisms (SNPs) in the VDR [8]. VDR polymorphisms such as *FokI*, *TaqI*, and *BsmI* affect the function of vitamin D, and they are evident in HNC [8]. Disruption of the VDR helps prevent the biological effects of vitamin D in oral cancer. In addition to VDR polymorphisms, advanced and aggressive oral cancers are characterized by low CYP27B1 and high CYP24A1 levels [14–17]. This favors oral cancer survival. The effects of VDR polymorphism, low CYP27B1, and high CYP24A1 indicate that vitamin D has significant and more informative prognostic value than clinicopathological parameters do. Some studies have supported the prognostic value of vitamin D because it is an indicator of the development of tumors in some tissues [9, 11, 18].

There is high expression of vitamin D deficiency in oral and extraoral malignant tumors [11, 18–21]. Oral cancer tissues show increased VDR expression [11]. This is a sign of increased demand for vitamin D. Several other studies have shown that the majority of patients with HNC have a higher incidence of vitamin D deficiency [8, 12, 22]. Bochen et al. [22], in a study involving 231 patients with HNSCC and 232 controls, reported a lower level of serum vitamin D in HNSCC patients. Vitamin D was associated with clinicopathological data. The lower serum levels of vitamin D in HNSCC patients were associated with positive lymph node status. In a systematic review by Pu et al. [8], HNC mortality increased with respect to insufficient serum levels of vitamin D. This finding indicated a direct relationship between HNC survival and serum vitamin D levels [8]. The anticancer effect of vitamin D is displayed by its ability to arrest or slow tumor growth and progression [13, 18, 21, 23], inhibit tumor angiogenesis, stimulate mutual adherence of cells, and enhance intercellular communication. It strengthens the inhibition of cellular proliferation [11, 13, 21]. A deficiency in vitamin D or a low level of vitamin D affects the anticancer effect of vitamin D, and this deficiency has been shown to be associated with increased cancer incidence, growth, resistance to chemotherapy, and poor outcomes [8].

The effects of 1,25$(OH)_2D_3$ on cancer cells were reported over three decades ago [24]. Two major effects were described at that time: the inhibition of proliferation and the enhancement of differentiation [18, 24]. An increase in 1,25 $(OH)_2D_3$ in melanoma patients was associated with the doubling time of melanoma cells [18]. Similar to leukemia cells, leukemic cells can differentiate into macrophages, indicating the lost malignant potential of leukemic cells [18]. Today, 1,25$(OH)_2D_3$ and its analogs slow the growth of cancer cells by arresting them in the G0/G1 phase of the cell cycle. It also induces their differentiation or apoptosis. It also alters cell adhesion and migration and reduces the invasiveness of cancer cells [18]. The supplementation of vitamin D in patients with advanced oral cancer undergoing chemotherapy significantly improved their quality of life and reduced the incidence of chemotherapy-induced toxicities such as oral mucositis, pain, and erythema [11].

Compared with the AJCC anatomic stages, the loss of VDR is present across stages, indicating increased malignant progression [11]. This finding adds value to the biological significance of the estimation of the VDR in the management of neoplasms and its prognostic value. Despite excellent reports on the benefit and significance of vitamin D in cancer, inconclusive differences have been reported, as reported by Kitchen and colleagues in a systematic review [11] and Fanidi et al. [2].

Metabolism of Vitamin D

Vitamin D_3 (cholecalciferol) is the natural form of vitamin D produced in the skin from the precursor 7-dehydrocholesterol [18, 25, 26]. The skin is the most important source of vitamin D [27]. Another source of vitamin D is diet. Foods such as dairy products, fish oils, and mushrooms are also sources of vitamin D [18, 25, 26]. The production of vitamin D by the skin depends on the exposure and intensity of ultraviolet irradiation. The availability of ultraviolet irradiation depends on the season and latitude [18, 25–27]. Irradiation of 7-dehydrocholesterol results in the production of previtamin D, which undergoes heterodimerization. Heterodimerization causes rearrangement of three double bonds, and vitamin D_3 is formed [16, 18, 26]. From the skin, vitamin D is transported in the blood to the liver, where it is bound to vitamin D binding protein (DBP), which is capable of binding vitamin D and its products [16, 18, 26]. In the liver, the hydroxylation of vitamin D at position 25 takes place to form 25-hydroxycholecalciferol (hydroxyvitamin D_3 [25$(OH)D_3$]) via the enzyme CYP2R1 (Fig. 1) [16, 18, 26].

Many cytochrome P-450 (CYP) enzymes are capable of converting vitamin D to 25$(OH)D_3$. These genes include CYP27A1 and CYP2D25 [16, 18, 26]. This process is not controlled by the concentration of the products. 25$(OH)D_3$ is transported from the liver to the kidney, where it is bound by DBP [27]. Megalin is a 600-kDa transmembrane protein that belongs to the low-density lipoprotein receptor superfamily. Megalin acts as a cell-surface receptor for DBP and leads to the uptake of 25$(OH)D_3$ by tubular epithelial cells through endocytic internalization [16, 18, 26]. A second surface cell receptor for DBP is cubulin at the proximal tubules. It is associated with Megalin and together they internalize complexes of DBP and 25$(OH)D_3$ [18, 26]. Disabled 2 (Dab 2) is another protein that serves as a cytoplasmic adaptor. Dab 2 works with Megalin to facilitate the cellular uptake of DBP/25$(OH)D_3$ through binding to the cytoplasmic tail of Megalin. This leads to proper routing of the receptor [18, 26]. 25$(OH)D_3$ is hydroxylated by renal 1α-hydroxylase (CYP27B1) in the proximal tubules at position 1 of the ring, leading to the formation of 1,25 $(OH)_2D_3$ (calcitriol), the active form of vitamin D [16, 23]. This protein is responsible for the biological actions of vitamin D in the human body through genomic and nongenomic pathways (Fig. 2) [18, 26]. This enzyme is present

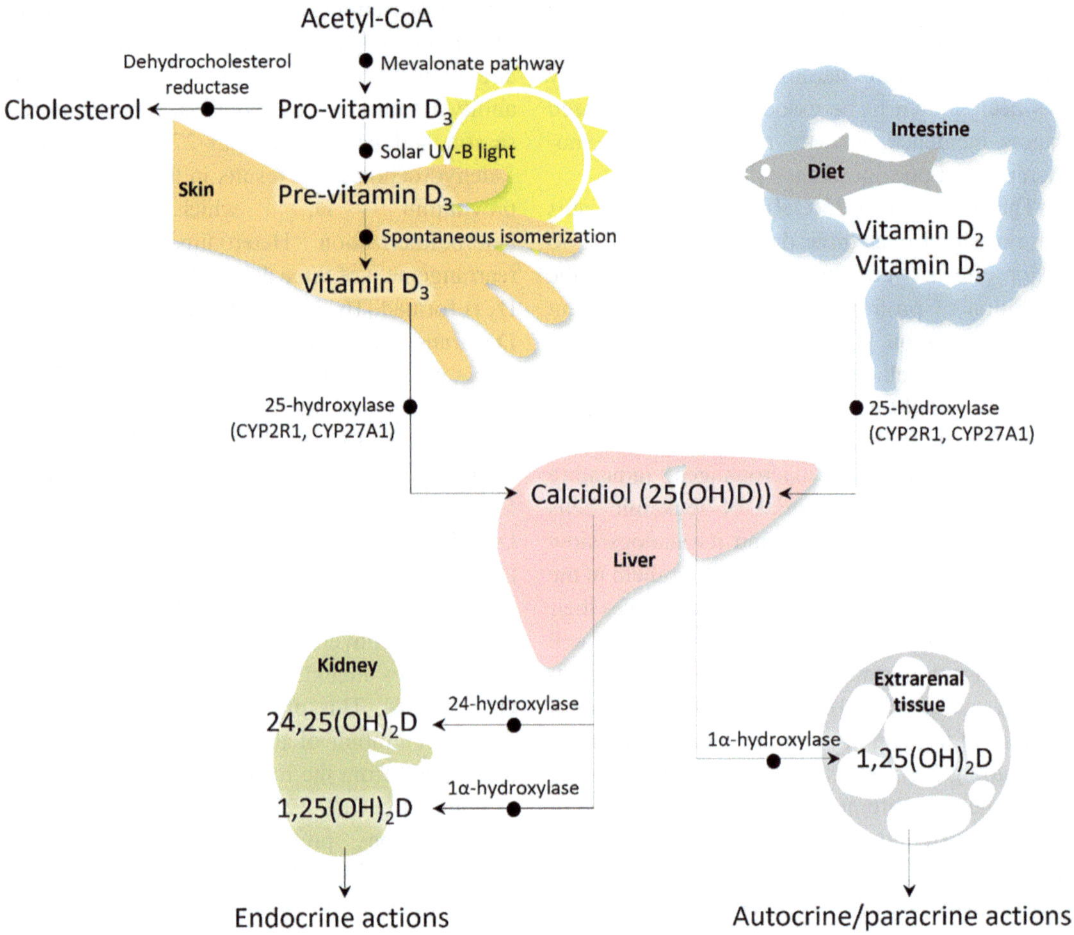

Fig. 1 Synthesis and activation of vitamin D

predominantly in the kidney and is the major contributor to the circulating level of 1,25 $(OH)_2D_3$ [18, 26]. The circulating concentration of 1,25$(OH)_2D_3$ controls this process.

Several human cells and some tissues express the enzyme CYP27B1, indicating the local production of 1,25$(OH)_2D_3$ in these cells [16, 19]. Tissues such as the placenta and cells such as defense cells, macrophages, monocytes, lymphocytes, and CD4+ and CD8+ T cells all express CYP27B. The expression of CYP27B1 has also been revealed in cancer cells, revealing the role of vitamin D in the development and suppression of cancer growth [16, 18, 27]. 25 $(OH)D_3$ in the kidney can also be converted into other products, such as 24,25$(OH)_2D_3$, through hydroxylation at C-24 by CYP24A1, which is a mitochondrial inner membrane cytochrome P-450 enzyme [18, 26]. The enzyme CYP24A1 limits the level of 1,25$(OH)_2D_3$ when the circulating level is elevated and converts it into 24,25$(OH)D_3$, leading to the formation of calcitroic acid and its excretion [18, 27]. This enzyme can also convert 1,25$(OH)D_3$, leading to the formation of 23-lactone [18]. All cells containing VDR also express CYP24A1 [18, 19]. This enzyme regulates the level of 1,25 $(OH)_2D_3$ within cells [18]. Elevated parathyroid hormone (PTH) usually results from hypocalcemia control of CYP27B1 and CYP24A1 in the kidney [18, 26]. The induction of CYP27B1 by PTH occurs indirectly through the nuclear orphan receptor 4A2 (NR4A2, also known as NUUR1). An elevated level of 1,25$(OH)_2D_3$ inhibits the

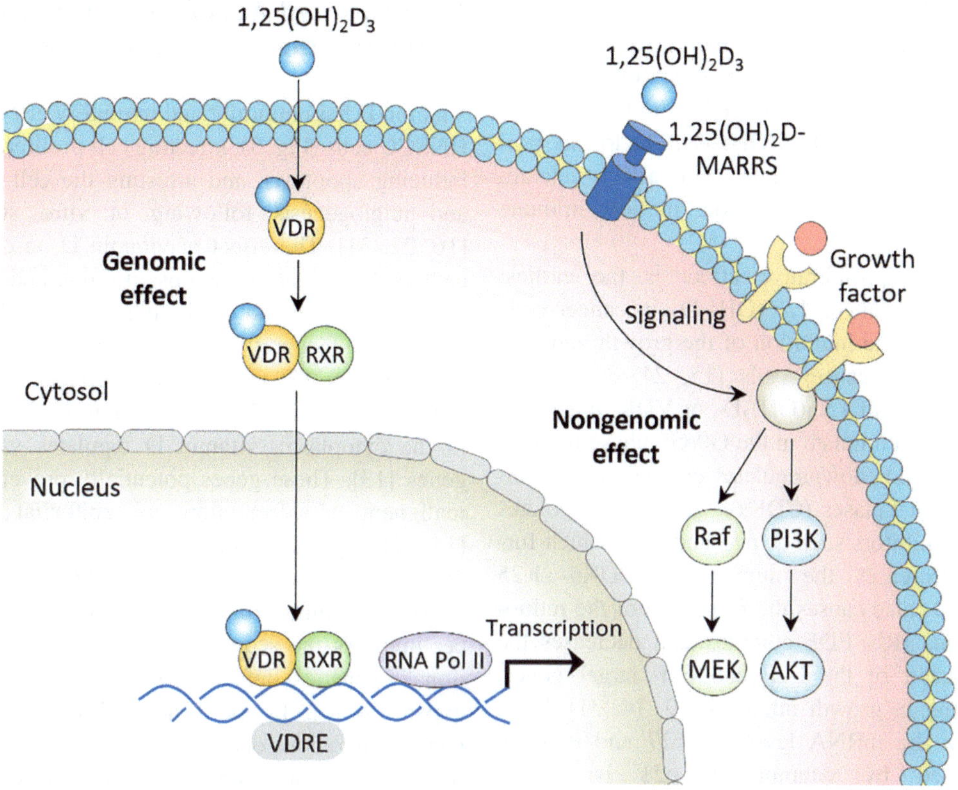

Fig. 2 Genomic and nongenomic effects of vitamin D

production of PTH [19]. Unlike the process of the formation of $1,25(OH)_2D_3$ in the kidney, the elevated level of $1,25(OH)_2D_3$ does not control the CYP27B1 produced by extrarenal cells such as immune cells [16, 18, 26].

General Mechanism of Action of Vitamin D in Cancers

There are two mechanisms of action of $1,25(OH)_2D_3$: genomic and nongenomic (Fig. 2) [14, 17, 28]. In terms of the genomic mechanism, $1,25(OH)_2D_3$ activated VDR/RXR, which bind directly to specific DNA sequences, such as vitamin D response elements (VDREs), in and around target genes, leading to either the activation or suppression of transcription [9, 17, 18, 27]. The heterodimerization of $1,25(OH)_2D_3$-VDR with RXR results in high-affinity binding of VDREs [18, 27].

Vitamin D improves the antitumor function of the immune system, and sufficient availability of vitamin D is necessary [22, 29]. The mechanism of action of the active form of vitamin D, $1,25(OH)_2D_3$, is mediated through vitamin D receptors (VDRs) [15, 18, 19]. VDRs are among the steroid receptor families and contain 427 amino acids with 2 core functional domains, the conserved NH2-terminal DNA binding domain (DBP) and the variable COOH-terminal ligand-binding domain (LBD) connected through the hinge region [18]. The VDR functions as an obligate heterodimer with RXR for the activation of vitamin D target genes [17, 18, 26]. The binding of $1,25(OH)_2D_3$ induces conformational changes that cause interactions with RXR, the complexes responsible for the activation of vitamin D target genes [17, 18].

Vitamin D regulates the expression of the VDR in cancer cells [27]. Most cancer cells express VDR, CYP27B1, and CYP24A1 [9, 14,

15, 17, 30]. The presence of these enzymes and receptors is an indication of their effects on cancer cell development and proliferation [9, 14, 15]. This provides cancer cells with the potential to locally regulate 1,25$(OH)_2D_3$. The locally produced 1,25$(OH)_2D_3$ does not affect calcium metabolism but plays a role in host immune responses [18].

The antiproliferative effect is the earliest observed event of 1,25$(OH)_2D_3$ in cancer cells because of the inhibition of the growth and proliferation of cancer cells [15, 24, 31]. When incubated with 1,25$(OH)_2D_3$ or VDR, most cancer cells accumulate in the G0/G1 phase because 1,25$(OH)_2D_3$ downregulates cyclins and cyclin-dependent kinases (CDKs), while it upregulates CDK inhibitors such as p21 and p27, which further decreases the number of CDKs. 1,25 $(OH)_2D_3$ also causes the formation of the retinoblastoma (Rb)-EDF complex and decreases the expression of the E2F and F2F target genes, resulting in growth inhibition [9, 14, 31]. Even though the mRNA levels of p27 and p53 are increased by vitamin D, p21 is mostly increased [9].

The effect of 1,25$(OH)_2D_3$ is not the same in different tumor cells (Fig. 3) [19]. In breast cancer, it induces C/EBPα and enhances VDR transcription, which is caused by C/EBPα [14]. This results in the inhibition of breast cancer cells. In human colon cancer, 1,25$(OH)_2D_3$ antagonizes Wnt/β-catenin signaling by inducing the expression of dickkopf 1, which inhibits Wnt, resulting in a reduction in cell proliferation [17, 18]. Treatment of cancer cells with 1,25$(OH)_2D_3$ decreased the interaction between β-catenin and T-cell factor (TCF) while enhancing the interaction between VDR and β-catenin [14]. It further enhances E-cadherin expression [14, 17], which causes the nuclear export of β-catenin and the induction of the expression of Dickkopf (DKK) 1, which is an extracellular Wnt inhibitor [14, 17]. These effects decrease the transcription of target genes, including c-myc, which is a key regulator of cell cycle progression [18].

Mechanism of Action and Prognostic Value of Vitamin D in Oral Cancers

Vitamin D has antitumor and inhibitory effects on HNSCC cells (Fig. 3), affecting cell proliferation, inducing apoptosis, and arresting the cell cycle and angiogenesis following in vitro studies [16, 22, 31]. The effect of vitamin D on cancer cells is dependent on the dose, timing, and duration of exposure and on tissue specific, lifestyle, and genetic factors [9, 31]. The presence of VDR is crucial for the antineoplastic effects of 1,25 $(OH)_2D_3$ [11, 13]. Through binding to the VDR in the cytoplasm, vitamin D regulates various genes [13]. These genes potentially prevent the malignant transformation of epithelial cells [13]. The epithelial cells can potentially convert 25-vitamin D into an active form [13].

The mechanism for the occurrence of oral cancer involves the sequential stimulation of additional genetic defects followed by clonal expansion [4]. In the head and neck region, genetic alterations occur due to oncogene activation and tumor suppressor gene inactivation [4, 32]. This interferes with the regulation of cell proliferation and programmed cell death [4, 32] (Fig. 3). The genetic alterations include gene amplification and the overexpression of oncogenes such as c-myc, erbB-2, epidermal growth factor (EGFR), and cyclin D1, as well as mutations, deletions, and hypermethylation, leading to p16 and p53 tumor suppressor gene inactivation [4]. Tumor suppressor genes and growth regulators are strongly affected [4]. Tumor suppressor genes and growth regulators are transducers of negative growth signals [4]. Genetic alterations involving the tumor suppressor genes p16 and p53 are common in oral cancers [4, 32]. Cell growth defects are caused when genetic abnormalities inactivate the p16 gene [4]. P16 is involved in cell cycle regulation, such as cell cycle arrest and apoptosis [4]. P53 maintains genomic stability [4]. P53 controls the cell cycle at the G1/S and GE/M checkpoints and induces apoptosis [4]. p53 gene inactivation has

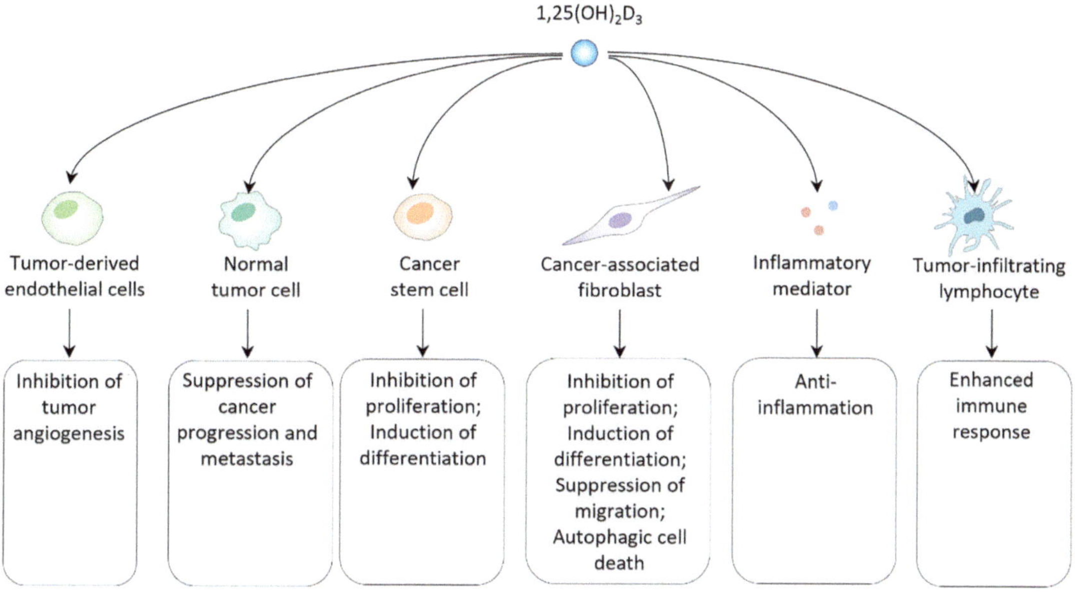

Fig. 3 Anticancer effects of vitamin D in tumor cells

been demonstrated to occur in oral squamous cell carcinoma via point mutations, deletions, and binding with cellular and viral proteins [4]. Suppression of Th1 cytokine genes increases tumor load and lymph node invasion [4]. Head and neck malignant tumor cells express interleukin-4 receptors in situ and interleukin-13, which are responsible for growth patterns and the active spread of tumor cells [4]. The multifactorial mechanism of oral carcinogenesis is illustrated in Fig. 4, in which epithelial cells are influenced by various genetic alterations (Fig. 4).

Vitamin D ($1,25(OH)_2D_3$) has anticancer effects in several ways (Fig. 5). Vitamin D causes the activation of bax and p-calpains, which are effective proapoptotic proteins that function to enhance the proapoptotic effect in different cancer cells, including oral cancer cells (Figs. 3 and 5) [9]. Vitamin D increases the mRNA expression of transforming growth factors. This is a potent antiproliferative cytokine in the early stage of cancer [9]. Vitamin D increases the expression of superoxide dismutase. This is responsible for the reduction in oxidative stress-induced DNA damage and loss of the DNA repair mechanism, which contributes to carcinogenesis and inflammatory cytokine production [9]. Vitamin D increases the expression of the cyclin-dependent kinase (CDK) inhibitors p21, RBL2, RBLP6, and forkhead box O (FOXO) tumor suppressors [13, 28, 31]. These functions counteract MAPK-mediated phosphorylation and growth [9, 13]. It also causes transcriptional repression of Bcl-2 and telomerase, which are pro-survival proteins [9]. In addition, it causes transcriptional CDK1 mRNA, which encodes the protein required for cell cycle progression [9], and suppresses vascular endothelial growth factor (VEGF), which is responsible for angiogenesis [9]. Vitamin D suppresses pro-inflammatory cyclooxygenase-2 [9]. Furthermore, it disrupts the function of β-catenin, which mediates Wnt signaling and is responsible for activating the transcription of genes encoding c-Myc and cyclin D1, hence controlling cell proliferation and insulin-like growth factor-stimulated tumor growth [9, 28].

In terms of cellular proliferation, it increases p21, p27, and p57 expression [13, 16, 28, 31]. p21, p27, and p57 are tumor suppressors that function through the inhibition of CDK2 activity. They are responsible for cell growth arrest to facilitate DNA damage repair [32]. An increase in p21, p27, and p57 results in a decrease in the expression of CDKs, cyclins, c-myc, and

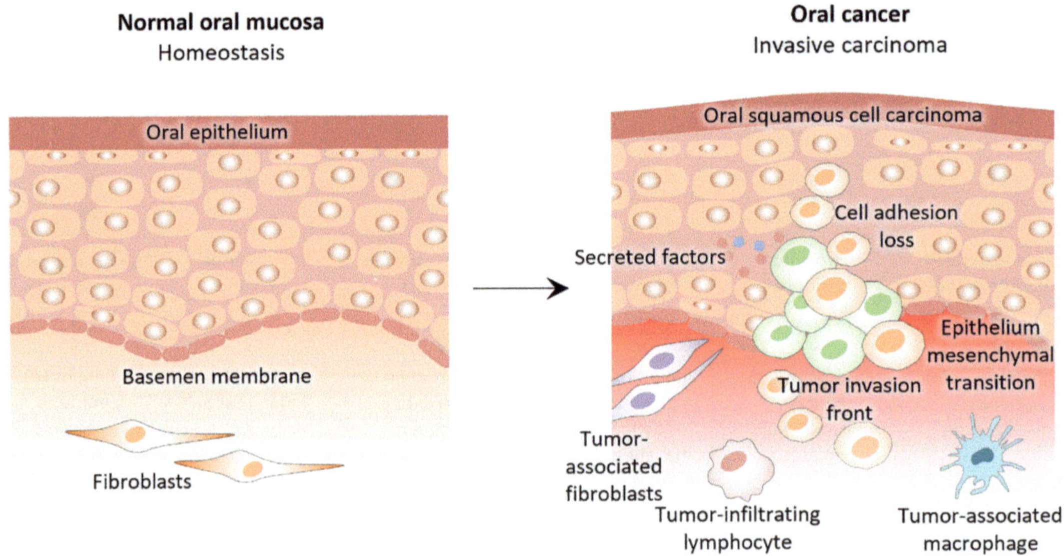

Fig. 4 Initiation of oral carcinogenesis

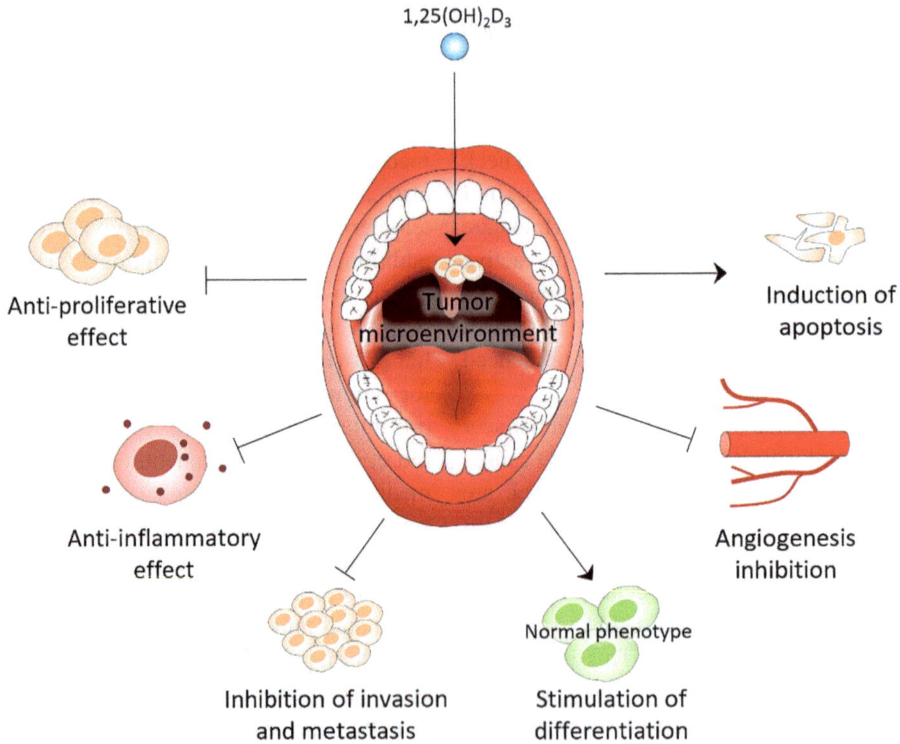

Fig. 5 Anticancer effects of vitamin D in oral cancer

RB [13, 16, 28]. In oral cancer, vitamin D downregulates LUCAT1, inhibiting tumor cell growth [33]. LUCAT1 works by signaling via the MAPK pathway to promote tumor cell proliferation. Therefore, vitamin D inhibits the LUCAT1-MAPK signaling pathway [33].

In apoptosis, $1,25(OH)_2D_3$ downregulates BCL-2 associated X (BAX), which is responsible for encoding the BCL2L4 protein, decreasing BCL-2 and increasing sensitivity to radiation and chemoradiotherapy. Under normal circumstances, upon activation, the BCL 2L4 protein heterodimerizes with Bcl2 and induces cell death [16, 28, 32]. Vitamin D and its analogs modulate key mediators of cancer cell apoptosis [32]. Specifically, they inhibit antiapoptotic proteins and/or stimulate proapoptotic proteins. For example, equitol, an epianalog of vitamin D, induces the mRNA expression of p27 and cyclin D1, which are involved in cell cycle arrest and apoptosis [32]. Caspases 8/10 and 3 are activated through the intrinsic pathway and promote apoptosis in squamous cell carcinoma cells by reducing the levels of antiapoptotic proteins, such as cellular inhibitor of apoptosis 1 (c-IAP1) and X-linked inhibitor of apoptotic proteins (XIAP) [32]. Both vitamin D and pecitol depend on VDR-mediated transcription of the CYP24A1 promoter to induce apoptotic effects [32]. Vitamin D promotes apoptosis by suppressing insulin-like growth factor 1 (IGF1) and activating forkhead box 03 [32].

Vitamin D increases the expression of differentiation factors, such as casein, lipids, PSA, and E-cadherin [16]. It inhibits the expression of COX 2 [16], which is responsible for the development and progression of head and neck cancer (Figs. 3 and 5) [34, 35]. It also inhibits PG receptor, stress kinase, and NF-kB signaling, increasing TIMP-1 and E-cadherin responses [16]. Low E-cadherin expression is associated with poor tumor differentiation, low survival, recurrence, and metastasis [36]. In terms of invasion, prognosis, and metastasis, vitamin D decreases the expression of MMP9, α6 integrin, and α4 integrins and plasminogen activator, and in angiogenesis, it decreases HIF-1α, VEGF, IL-8, tenascin C, and PGE2 levels [16, 32, 37]. Bochen et al. [22] analyzed the antitumor activity of vitamin D in HNSCC. They performed immunohistochemical staining of the primary tumor using CD3 as a marker to detect cells (using CD3 as a marker), such as helper T cells (CD4), cytotoxic T cells (CD8), natural killer cells (CD56), macrophages (CD68), M1 macrophages (CD11c), and M2 macrophages (CD163), to analyze the intra- and peritumoral immune cell infiltration in patients with higher vitamin D serum levels and lower serum levels. They reported significantly greater intramural and/or stromal infiltration of CD3+ T cells, helper T cells, cytotoxic T cells, NK cells, CD68+ macrophages, and M1 macrophages in vitamin D-high patients than in vitamin D-low patients. M2 macrophage staining revealed a lower tendency toward lower peritumoral infiltration in vitamin D-high patients. Anand et al. [11] reported improved quality of life following supplementation with vitamin D in patients with advanced inoperable oral cancer. The quality of life was noted in terms of improved swallowing, the clinical presentation of the lesions, and the quality of life on a four-point Likert scale. Fanidi et al. [2] assessed the association between vitamin D and HNC in 497 patients with HNC and reported that a decreased risk of HNC in patients with a high level of vitamin D and doubled serum vitamin D was associated with a 45% lower risk for HNC.

The prognostic value of vitamin D has been revealed by Bochen and colleagues [22]. Bochen et al. [22] reported shorter overall survival in patients with low vitamin D serum levels. Bochen et al. [22] compared the HPV-positive tumor status and serum vitamin D levels in HNSCC patients. HPV-positive status has been reported to be a significant predictor of better overall survival. Their study did not find any correlation of vitamin D serum levels with HPV-positive tumor status; however, they concluded that this correlation could result from the small sample size of only 35 patients who were HPV-positive. Surprisingly, low serum vitamin D levels were correlated with shorter overall survival in HPV-negative patients.

The presence of CYP24A1 in tumor cells counteracts the effect of $1,25(OH)_2D_3$. CYP24A1 is highly expressed in highly malignant and metastatic tumors [14–17]. The expression of this enzyme is associated with increased resistance of tumors to the effect of $1,25(OH)_2D_3$ [7, 14–16]. Although there is little information on this effect in oral cancer, it has been revealed in breast and colorectal tumors. The presence of CYP24A1 is also associated with poor survival in cancer patients. This has been observed in lung cancer, where less survival is observed in patients whose tumors express higher levels of CYP24A1 [14]. Poorly differentiated tumors display less CYP27B1 protein than well-differentiated tumors do [14, 17, 23]. With the few studies reporting on this association, no conclusion has been drawn regarding the association between CYP27B1 and disease progression. Two different views exist: While some studies support this association, few others refute the role of $1{,}25(OH)_2D_3$ in tumor progression [7, 17, 23, 28, 31]. Among the factors associated with a lack of association are short follow-up time, duration of vitamin D supplementation, lifestyle, ethnicity, environmental living conditions, and variation in geographical location [18]. This finding calls for continued exploration and documentation of the role of $1{,}25(OH)_2D_3$ in oral cancer, which still presents challenges for identifying promising prognostic parameters. More evidence has been provided to support an inverse relationship between colorectal and fatal breast cancers compared with oral cancers [16, 17, 23].

Oropharyngeal cancer cells may interfere with the effects of vitamin D by affecting its metabolism through changing its availability and binding to the VDR [7] and polymorphisms of the VDR [13]. This effect is achieved by cancer cells exhibiting a reduction in VDR expression [7, 13]. Cancer cells exhibit Ras activation, which impairs vitamin D-mediated transcription. There is high expression of CYP27B1 and CYP24A1, enzymes responsible for the degradation of vitamin D metabolites [7, 9]. The expression of CYP24A1 is greater in oral cancers that are resistant to the action of $1{,}25(OH)_2D_3$ [7, 16, 17]. This has been observed more often in cancers that have poor clinical outcomes, such as poorly differentiated and aggressive cancers [9, 16, 17], indicating that high expression of CYP24A1 is associated with poor prognosis [7, 14, 16, 17]. The use of CYP24A1 inhibitors has been shown to reverse the sensitivity of cancers to the utilization of $1{,}25(OH)_2D_3$ [14, 16]. Furthermore, the use of cytochrome P450 inhibitors favors the biological actions of $1{,}25(OH)_2D_3$ [16].

The physiological activity of CYP24A1 and CYP27B1 in cancer cells has promising prognostic value [13, 15]. The use of $1{,}25(OH)_2D_3$ in the early stages may reduce the biological activity of most cancer cells [14]. The regulation of CYP27B1 in cancer cells varies with the tissue and stage of the tumor [16]. The degree of cellular differentiation affects the expression of CYP27B1 because it is more highly expressed in well-differentiated tumors than in poorly differentiated tumors [14, 15]. The opposite is true for CYP24A1, which is more highly expressed in poorly differentiated tumors than in well-differentiated tumors [14, 15]. While CYP27B1 is associated with the effective biological effect of $1{,}25(OH)_2D_3$ on tumor cells, CYP24A1 is associated with cancer cell resistance to the utilization of $1{,}25(OH)_2D_3$ and hence a poor prognosis [14–16].

Conclusion

There is substantial evidence for the direct biological activity of vitamin D in oral cancer and its ability to predict progression, apoptosis, recurrence, response to therapy, and prognosis. The presence of the VDR and CYP27B1 enzymes signifies the role of vitamin D in oral cancer cells. The presence of the CYP24A1 enzyme and mutations in VDRs are evidence of resistance to the utilization of vitamin D in aggressive cancer cells, which affects tumor suppressor gene activity and apoptosis. Substantial evidence shows that low levels of vitamin D in oral cancer patients are associated with short survival, poor differentiation, resistance to irradiation, and the occurrence of recurrence. The outcomes of these effects include tumor growth, spread, and progression. These findings call for the comparison of vitamin D with other clinicopathological parameters to explore its prognostic value in oral cancers at any stage.

Recommendation

The serum levels of vitamin D, VDR, CYP27B1, and CYP24A1 may be useful indicators of the biological activity of tumor cells and hence predictors of tumor progression, sensitivity to treatment, and patient prognosis in patients with HNC. Substantial evidence from the literature on the role of vitamin D in the biological activity of oral cancer cells calls for research to correlate clinicopathological prognostic parameters with vitamin D, VDR, CYP27B1, and CYP24A1. Further research should consider the associations between vitamin D and survival and treatment outcomes in patients with HNC. The quantification of vitamin D, VDR, CYP27B1, and CYP24A1 promises to improve the management of oral cancer lesions at all stages and the determination of patient prognosis.

Conflict of Interest Declaration The authors declare that they have no conflicts of interest.

References

1. Johnson DE, Burtness B, Leemans CR, Lui VWY, Bauman JE, Grandis JR (2020) Head and neck squamous cell carcinoma. Nat Rev Dis Prim 6:92. https://doi.org/10.1038/s41572-020-00224-3
2. Fanidi A, Muller DC, Midttun Ø, Ueland PM, Vollset SE, Relton C, Vineis P, Weiderpass E, Skeie G, Brustad M, Palli D, Tumino R, Grioni S, Sacerdote C, Bueno-De-Mesquita HB, Peeters PH, Boutron-Ruault MC, Kvaskoff M, Cadeau C, Huerta JM, Sánchez MJ, Agudo A, Lasheras C, Quirós JR, Chamosa S, Riboli E, Travis RC, Ward H, Murphy N, Khaw KT, Trichopoulou A, Lagiou P, Papatesta EM, Boeing H, Kuehn T, Katzke V, Steffen A, Johansson A, Brennan P, Johansson M (2016) Circulating vitamin D in relation to cancer incidence and survival of the head and neck and oesophagus in the EPIC cohort. Sci Rep 6:1–11. https://doi.org/10.1038/srep36017
3. Anwar N, Pervez S, Chundriger Q, Awan S, Moatter T, Ali TS (2020) Oral cancer: clinicopathological features and associated risk factors in a high risk population presenting to a major tertiary care center in Pakistan. PLoS One 15:1–15. https://doi.org/10.1371/journal.pone.0236359
4. Mehrotra R, Yadav S (2006) Oral squamous cell carcinoma: etiology, pathogenesis and prognostic value of genomic alterations. Indian J Cancer 43:60–66. https://doi.org/10.4103/0019-509X.25886
5. Taghavi N, Yazdi I (2015) Prognostic factors of survival rate in oral squamous cell carcinoma: clinical, histologic, genetic and molecular concepts. Arch Iran Med 18:314–319
6. Ellington TD, Henley SJ, Senkomago V, O'Neil ME, Wilson RJ, Singh S, Thomas CC, Wu M, Richardson LC (2020) Trends in incidence of cancers of the oral cavity and pharynx-United States 2007-2016. MMWR Morb Mortal Wkly Rep 69:433–438. https://www.cdc.gov/cancer/dataviz
7. Grimm M, Cetindis M, Biegner T, Lehman M, Munz A, Teriete P, Reinert S (2015) Serum vitamin D levels of patients with oral squamous cell carcinoma (OSCC) and expression of vitamin D receptor in oral precancerous lesions and OSCC. Med Oral Patol Oral Cir Bucal 20:e188–e195. https://doi.org/10.4317/medoral.20368
8. Pu Y, Zhu G, Xu Y, Zheng S, Tang B, Huang H, Wu IXY, Huang D, Liu Y, Zhang X (2021) Association between vitamin D exposure and head and neck cancer: a systematic review with meta-analysis. Front Immunol 12:1–11. https://doi.org/10.3389/fimmu.2021.627226
9. Kingsley K, Bergman C, Keiserman M, Mobley C (2013) Oral cancer risk and vitamin D status, intake and supplementation: a review critical review oral cancer risk and vitamin D status, intake and – supplementation: a review. Open Access Cancer 1:1–5. https://digitalscholarship.unlv.edu/dental_fac_articles/52
10. Baniebrahimi G, Mir F, Khanmohammadi R (2020) Cancer stem cells and oral cancer: insights into molecular mechanisms and therapeutic approaches. Cancer Cell Int 20:1–15. https://doi.org/10.1186/s12935-020-01192-0
11. Anand A, Singh S, Sonkar AA, Husain N, Singh KR, Singh S, Kushwaha JK (2017) Expression of vitamin D receptor and vitamin D status in patients with oral neoplasms and effect of vitamin D supplementation on quality of life in advanced cancer treatment. Wspolczesna Onkol 21:145–151. https://doi.org/10.5114/wo.2017.68623
12. Udeabor SE, Albejadi AM, Al-Shehri WAK, Onwuka CI, Al-Fathani SY, Al Nazeh AA, Aldhahri SF, Alshahrani FA (2020) Serum levels of 25-hydroxyvitamin D in patients with oral squamous cell carcinoma: making a case for chemoprevention. Clin Exp Dent Res 6:428–432. https://doi.org/10.1002/cre2.294
13. Hama T, Norizoe C, Suga H, Mimura T, Kato T, Moriyama H, Urashima M (2011) Prognostic significance of vitamin D receptor polymorphisms in head and neck squamous cell carcinoma. PLoS One 6:8–13. https://doi.org/10.1371/journal.pone.0029634
14. Carlberg C, Muñoz A (2022) An update on vitamin D signaling and cancer. Semin Cancer Biol 79:217–230. https://doi.org/10.1016/j.semcancer.2020.05.018

15. Dou R, Ng K, Giovannucci EL, Manson JE, Qian ZR, Ogino S (2016) Vitamin D and colorectal cancer: molecular, epidemiological and clinical evidence. Br J Nutr 115:1643–1660. https://doi.org/10.1017/S0007114516000696
16. Beatrice JE (2015) Anticancer effects of vitamins. Am J Hematol Oncol 11:26–36. https://www.gotoper.com/publications/ajho/2015/2015oct/anticancer-effects-of-vitamin-d
17. Khammissa RAG, Fourie J, Motswaledi MH, Ballyram R, Lemmer J, Feller L (2018) The biological activities of vitamin D and its receptor in relation to calcium and bone homeostasis, cancer, immune and cardiovascular systems, skin biology, and oral health. Biomed Res Int 2018:9276380. https://doi.org/10.1155/2018/9276380
18. Christakos S, Dhawan P, Verstuyf A, Verlinden L, Carmeliet G (2015) Vitamin D: metabolism, molecular mechanism of action, and pleiotropic effects. Physiol Rev 96:365–408. https://doi.org/10.1152/physrev.00014.2015
19. Vuolo L, Di Somma C, Faggiano A, Colao A (2012) Vitamin D and cancer. Front Endocrinol (Lausanne) 3:1–13. https://doi.org/10.3389/fendo.2012.00058
20. International Agency for Research on Cancer (2010) Vitamin D and cancer. In: Bioactive foods and extracts: cancer treatment and prevention. CRC Press, Boca Raton, pp 481–495. https://doi.org/10.1201/b10330
21. Imran S, Shah A (2020) Beneficial role of vitamin d in common cancers: is the evidence compelling enough? WCRJ 7:1–9. https://doi.org/10.32113/wcrj_20205_1574
22. Bochen F, Balensiefer B, Körner S, Bittenbring JT, Neumann F, Koch A, Bumm K, Marx A, Wemmert S, Papaspyrou G, Zuschlag D, Kühn JP, Al Kadah B, Schick B, Linxweiler M (2018) Vitamin D deficiency in head and neck cancer patients–prevalence, prognostic value and impact on immune function. Onco Targets Ther 7:1–10. https://doi.org/10.1080/2162402X.2018.1476817
23. Garland CF, Garland FC, Gorham ED, Lipkin M, Newmark H, Mohr SB, Holick MF (2006) The role of vitamin D in cancer prevention. Am J Public Health 96:252–261. https://doi.org/10.2105/AJPH.2004.045260
24. Holick MF (2004) Vitamin D: importance in the prevention of cancers, type 1 diabetes, heart disease, and osteoporosis. Am J Clin Nutr 79:362–371. https://doi.org/10.1093/ajcn/79.3.362
25. Akimbekov NS, Ortoski RA, Razzaque MS (2020) Effects of sunlight exposure and vitamin D supplementation on HIV patients. J Steroid Biochem Mol Biol 200:105664. https://doi.org/10.1016/j.jsbmb.2020.105664
26. Mumena CH, Mudhihiri MH, Sasi R, Mlawa M, Nyerembe S, Akimbekov NS, Razzaque MS (2021) The relevance of vitamin D in the oral health of HIV infected patients. J Steroid Biochem Mol Biol 211:105905. https://doi.org/10.1016/j.jsbmb.2021.105905
27. Mechanisms M, Implications T (2020) Vitamin D signaling in inflammation and cancer: molecular mechanisms and therapeutic implications. Molecules 25:1–31. https://doi.org/10.3390/molecules25143219
28. Negri M, Gentile A, de Angelis C, Montò T, Patalano R, Colao A, Pivonello R, Pivonello C (2020) Vitamin D-induced molecular mechanisms to potentiate cancer therapy and to reverse drug-resistance in cancer cells. Nutrients 12:1–25. https://doi.org/10.3390/nu12061798
29. Campbell MJ, Trump DL (2017) Vitamin D receptor signaling and cancer. Endocrinol Metab Clin N Am 46:1009–1038. https://doi.org/10.1016/j.ecl.2017.07.007
30. Kingsley K, Bergman C, Keiserman M, Mobley C (2013) Oral cancer risk and vitamin D status, intake and supplementation: a review. Open Access Cancer 1:1–5. https://doi.org/10.13172/2053-3918-1-1-545
31. Mäkitie A, Tuokkola I, Laurell G, Mäkitie O, Olsen K, Takes RP, Florek E, Szyfter K, Sier CFM, Ferlito A (2021) Vitamin D in head and neck cancer: a systematic review. Curr Oncol Rep 23:5. https://doi.org/10.1007/s11912-020-00996-7
32. Walia C, Roy S (2020) The emerging role of vitamin D in oral epithelial pathology: a review. J Dent Oral Disord Ther 8:1–7. www.symbiosisonline.org, www.symbiosisonlinepublishing.com
33. Jin T, Guo Y, Huang Z, Zhang Q, Huang Z, Zhang Y, Huang Z (2020) Vitamin D inhibits the proliferation of Oral Squamous Cell Carcinoma by suppressing lncRNA LUCAT1 through the MAPK pathway. J Cancer 11:5971–5981. https://doi.org/10.7150/jca.45389
34. Frejborg E, Salo T, Salem A (2020) Role of cyclooxygenase-2 in head and neck tumorigenesis. Int J Mol Sci 21:1–17. https://doi.org/10.3390/ijms21239246
35. Li TJ, Cui J (2013) COX-2, MMP-7 expression in oral lichen planus and oral squamous cell carcinoma. Asian Pac J Trop Med 6:640–643. https://doi.org/10.1016/S1995-7645(13)60110-8
36. López-Verdín S, Martínez-Fierro MDLL, Garza-Veloz II, Zamora-Perez A, Grajeda-Cruz J, González-González R, Molina-Frechero N, Arocena-Sutz M, Bologna-Molina R (2019) E-Cadherin gene expression in oral cancer: clinical and prospective data. Med Oral Patol Oral Cir Bucal 24:e444–e451. https://doi.org/10.4317/medoral.23029
37. Santosh A, Jones T, Harvey J (2016) A review on oral cancer biomarkers: understanding the past and learning from the present. J Cancer Res Ther 12:486–492. https://doi.org/10.4103/0973-1482.176414

Glucagon-Like Peptide-1: The Role of Calcium in Gut-Glucose Axis

Astrid-Ines Foamkom, Hosam G. Abdelhady, and Mohammed S. Razzaque

Abstract

Glucagon-like peptide-1 (GLP-1) is a key metabolic hormone secreted by intestinal L-cells in response to nutrient ingestion. It plays a central role in regulating glucose homeostasis by enhancing insulin secretion, inhibiting glucagon release, delaying gastric emptying, and suppressing appetite. While glucose, fat, and protein are established stimulators of GLP-1 release, calcium has emerged as a key modulator of this enteroendocrine axis. Recent studies have shown that extracellular calcium, acting through the calcium-sensing receptor (CaSR) and voltage-gated calcium channels, triggers intracellular calcium influx, which is essential for GLP-1 exocytosis from L-cells. This highlights the role of calcium not merely as a dietary mineral but as a signaling ion that links nutrient sensing in the gastrointestinal tract to the hormonal regulation of blood glucose homeostasis. This chapter briefly explains how calcium influences the GLP-1-mediated gut-glucose axis.

A.-I. Foamkom · M. S. Razzaque (✉)
Department of Medical Education, School of Medicine, University of Texas Rio Grande Valley (UTRGV), Edinburg, TX, USA
e-mail: mohammed.razzaque@utrgv.edu

H. G. Abdelhady
Department of Physiology & Pharmacology, College of Osteopathic Medicine, Sam Houston State University, Conroe, TX, USA

Keywords

Glucagon-like peptide-1 · GLP-1 receptor agonists · Hypoglycemia · Gastric emptying · Calcium · Diabetes mellitus

GLP-1 Physiology and Systemic Functions

GLP-1 is synthesized from the proglucagon gene, which undergoes tissue-specific posttranslational processing. In pancreatic α-cells, this gene encodes glucagon, whereas in intestinal L-cells and certain neurons, it yields GLP-1 and GLP-2 [1–3]. While GLP-2 primarily promotes intestinal growth and nutrient absorption, GLP-1 exerts wide-ranging metabolic effects. These include stimulating glucose-dependent insulin secretion from pancreatic β-cells, reducing glucagon release, delaying gastric emptying, decreasing intestinal motility, and promoting β-cell survival (Fig. 1) [4]. GLP-1 also acts through local receptors on enteric neurons and vagal afferents, modulating gut motility, appetite, and pancreatic exocrine function [5].

Although GLP-1 has a short half-life due to rapid degradation by dipeptidyl peptidase-4 (DPP-4), its receptor is widely expressed in the pancreas, gut epithelium, immune cells, kidneys, lungs, and brain [6, 7]. This broad distribution underpins its diverse physiological roles, from modulating glucose absorption and

M. S. Razzaque (ed.), *Vitamin D Function*, Advances in Experimental Medicine and Biology 1493,
https://doi.org/10.1007/978-3-032-04357-3_10

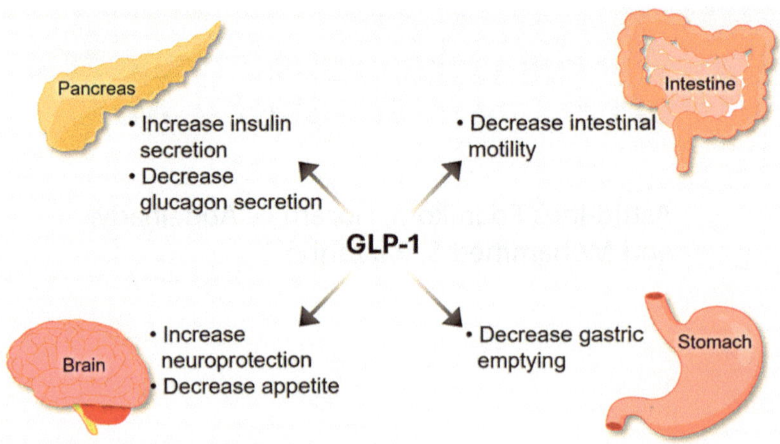

Fig. 1 The functions of GLP-1 in the stomach, intestine, pancreas, and brain are mostly mediated by direct interactions with GLP-1 receptors in these organs to affect glucose homeostasis

gastrointestinal motility to contributing to cardiovascular and neuroendocrine regulation [8, 9].

GLP-1 Secretion and the Role of Calcium

GLP-1 secretion from L-cells is highly dependent on calcium signaling [10]. The CaSR, expressed on L-cell membranes, responds to extracellular calcium as well as amino acids and peptides, especially those derived from protein digestion [11]. Co-ingestion of dietary protein, especially whey protein, and calcium has been shown to significantly potentiate GLP-1 secretion, implicating nutrient-nutrient interactions at the level of CaSR activation [10, 11]. Beyond its role in L-cells, GLP-1 also enhances intracellular calcium mobilization in pancreatic β-cells, leading to elevated cytosolic calcium levels that trigger insulin exocytosis [12]. Nutritionally, combining whey protein with milk-derived calcium minerals is a promising strategy to amplify GLP-1 secretion [11]. Proposed mechanisms by which calcium may augment GLP-1 secretion include (1) direct activation of CaSR on L-cells, (2) enhancement of L-cell sensitivity to other luminal nutrients, (3) modulation of intracellular calcium-dependent signaling pathways, and (4) delayed intestinal calcium absorption, especially in the distal ileum, where L-cell density is highest. Of note, milk-derived calcium salts, such as calcium phosphate, exhibit slower absorption kinetics than soluble salts like calcium citrate. This delayed absorption extends calcium availability in the ileum, enhancing GLP-1 secretion at distal gut sites.

Nutritional Modulation and Therapeutic Potentials of GLP-1

Studies have found that a 3-week intervention with calcium phosphate significantly increased postprandial GLP-1 concentrations compared to placebo [10, 11]. Notably, co-ingestion of whey protein hydrolysate with milk (a rich calcium source) elicited some of the highest postprandial GLP-1 levels ever reported. In one study, this combination increased GLP-1 secretion by ~25% compared to whey protein alone, suggesting a synergistic effect [10, 11]. While nondairy calcium sources can also enhance GLP-1 secretion, they appear to be less effective, particularly without co-ingested protein. These findings support the development of cost-effective, food-based strategies for augmenting endogenous GLP-1 secretion to aid in appetite regulation and glycemic control.

Metabolic diseases such as type 2 diabetes mellitus (T2DM) and obesity are complex, multi-system disorders influenced by a combination of genetic, nutritional, environmental, and behavioral factors [13–16]. Recently, GLP-1 receptor agonists (GLP-1 RAs) have emerged as effective treatments for T2DM and obesity [17]. GLP-1 RAs mimic the action of endogenous GLP-1, which is secreted in response to nutrient intake. These agents enhance glucose-dependent insulin secretion from pancreatic β-cells, suppress glucagon release from α-cells, and reduce hepatic glucose output. Clinically, they have been shown to lower blood glucose levels and HbA1c by approximately 0.8–1.5%, depending on the specific agent and dosage [18, 19]. Beyond glycemic control, GLP-1 RAs exhibit cardioprotective effects, reducing adverse cardiovascular events in T2DM patients [20]. Despite their benefits, GLP-1 RAs are generally prescribed as second- or third-line therapies, often in combination with first-line agents such as metformin. Their use may be limited by gastrointestinal side effects, including nausea, gastroparesis, pancreatitis, and rare cases of intestinal obstruction [21]. Consequently, new-generation agents such as tirzepatide, a dual GLP-1/glucose-dependent insulinotropic polypeptide (GIP) agonist, and orforglipron, an oral GLP-1 RA, offer improved efficacy, bioavailability, and delivery options [22].

However, gastrointestinal side effects (e.g., nausea, gastroparesis, and rare intestinal obstruction) and long-term safety remain considerations [21]. GLP-1's pleiotropic effects extend to reducing blood pressure, improving endothelial function, enhancing energy expenditure, and suppressing inflammation [23, 24]. For instance, a phase II clinical trial of lixisenatide, a GLP-1 RA, has shown promise in slowing the progression of motor disability in early Parkinson's disease [25, 26]. Emerging research has also explored multifunctional GLP-1-based molecules. One such compound, AMG 133, is a bispecific molecule combining a GIP receptor antagonist antibody with two GLP-1 agonist moieties. Preclinical studies have demonstrated significant weight loss and improved metabolic parameters in animal models [27]. Similarly, a bimodal molecule integrating NMDA receptor antagonism with GLP-1 agonism has been reported to reverse obesity, hyperglycemia, and dyslipidemia in rodent models of metabolic syndrome [28]. Together, these findings underscore the expanding therapeutic scope of GLP-1, extending far beyond glycemic control to encompass cardiovascular, neurological, hepatic, and metabolic health.

Therapeutic Limitations of GLP-1

Despite their benefits, GLP-1 RAs may impair counterregulatory responses to hypoglycemia. Studies in diabetic mouse models reveal that excessive intestinal GLP-1 and receptor expression suppress glucagon, epinephrine, and norepinephrine release during hypoglycemia, compromising glucose recovery. Physiologically, GLP-1 suppresses α-cell glucagon release under hyperglycemia. However, sustained receptor stimulation may blunt glucagon secretion even during hypoglycemia. Chronic GLP-1 RA exposure may also impair islet and central glucose sensing, delaying activation of sympathetic responses and increasing the risk of hypoglycemia-associated autonomic failure (HAAF), a condition marked by reduced awareness of hypoglycemia and diminished hormonal counterregulation [29, 30]. This syndrome is particularly concerning individuals receiving intensive glucose-lowering therapy, where frequent or unrecognized hypoglycemic episodes may exacerbate autonomic dysfunction and elevate the risk of severe hypoglycemia and cardiovascular complications [31]. However, the relationship between GLP-1 signaling and impaired counterregulation remains complex and context-dependent. Some studies suggest that short-term exposure to GLP-1 RAs does not significantly affect hormonal responses to hypoglycemia [32], indicating that duration of exposure, glycemic context, or coexisting pathophysiological factors may modulate this effect. In clinical practice, the use of GLP-1 Ras, particularly in patients with type 1 diabetes or those prone to frequent hypoglycemia, requires careful monitoring. Risk-benefit assessments should consider both the glycemic and non-glycemic advantages of GLP-1 analogs and the potential risk of impairing counterregulatory mechanisms, especially under intensive insulin regimens or in settings of recurrent hypoglycemia.

Conclusion

GLP-1 receptor agonists represent a major advancement in the treatment of type 2 diabetes and obesity due to their ability to enhance glucose-dependent insulin secretion, suppress appetite, and promote weight loss [33]. However, accumulating evidence suggests that excessive or chronic stimulation of intestinal GLP-1 pathways may impair physiological counterregulatory responses to hypoglycemia. This includes suppression of glucagon, epinephrine, and norepinephrine release, potentially increasing the risk of hypoglycemia-associated autonomic failure (HAAF), particularly in insulin-dependent individuals. These findings highlight the importance of patient-specific risk assessment when initiating or intensifying GLP-1-based therapies, especially in those prone to recurrent hypoglycemia.

Comparative research on different GLP-1 receptor agonists, including dual agonists like tirzepatide and oral non-peptide agents like orforglipron, is warranted to elucidate variations in safety profiles, efficacy, and risk of counterregulatory impairment. Additionally, prolonged GLP-1 exposure may alter central and peripheral glucose sensing mechanisms, underscoring the need for longitudinal studies to assess neuroendocrine and autonomic adaptations [34]. Beyond pharmacotherapy, integrating GLP-1-based drugs with nutrition and lifestyle strategies is essential for sustainable outcomes. As Mozaffarian emphasized, reliance solely on costly medications is insufficient to address the obesity epidemic. Nutritional interventions, such as combining calcium and protein to enhance endogenous GLP-1, offer affordable and physiologically sound alternatives. Finally, expanding indications for GLP-1 therapies in cardiovascular, hepatic, renal, and neurological diseases should be explored through rigorous clinical trials. Future efforts must balance efficacy, safety, affordability, and equity to maximize the societal impact of GLP-1-targeted strategies [35].

Acknowledgments We want to express our sincere gratitude to Dr. Nuraly Akimbekov, PhD (Al-Farabi Kazakh National University, Kazakhstan) for drawing the illustration. A.F. is a Doctor of Medicine (MD) student at the University of Texas Rio Grande Valley (UTRGV), Edinburg, Texas, USA. Some of the information has been collected from online sources, including OpenAI GPT-3.5 and Google Scholar. The authors are solely responsible for the content of the chapter. M.S.R. is a visiting professor at the University Kuala Lumpur Royal College of Medicine Perak, Ipoh, Malaysia.

References

1. Drucker DJ (2007) The role of gut hormones in glucose homeostasis. J Clin Invest 117:24–32. https://doi.org/10.1172/jci30076
2. Mariam Z, Niazi SK (2024) Glucagon-like peptide agonists: a prospective review. Endocrinol Diabetes Metab 7:e462. https://doi.org/10.1002/edm2.462
3. Drucker DJ, Yusta B (2014) Physiology and pharmacology of the enteroendocrine hormone glucagon-like peptide-2. Annu Rev Physiol 76:561–583. https://doi.org/10.1146/annurev-physiol-021113-170317
4. Andersen A et al (2018) Glucagon-like peptide 1 in health and disease. Nat Rev Endocrinol 14:390–403. https://doi.org/10.1038/s41574-018-0016-2
5. Halim MA et al (2018) Glucagon-like peptide-1 inhibits prandial gastrointestinal motility through myenteric neuronal mechanisms in humans. J Clin Endocrinol Metab 103:575–585. https://doi.org/10.1210/jc.2017-02006
6. Gilbert MP, Pratley RE (2020) GLP-1 analogs and DPP-4 inhibitors in type 2 diabetes therapy: review of head-to-head clinical trials. Front Endocrinol (Lausanne) 11:178. https://doi.org/10.3389/fendo.2020.00178
7. Carlessi R et al (2017) GLP-1 receptor signalling promotes beta-cell glucose metabolism via mTOR-dependent HIF-1alpha activation. Sci Rep 7:2661. https://doi.org/10.1038/s41598-017-02838-2
8. Holst JJ et al (2022) Actions of glucagon-like peptide-1 receptor ligands in the gut. Br J Pharmacol 179:727–742. https://doi.org/10.1111/bph.15611

9. Zhao X et al (2021) GLP-1 receptor agonists: beyond their pancreatic effects. Front Endocrinol (Lausanne) 12:721135. https://doi.org/10.3389/fendo.2021.721135
10. Watkins JD et al (2021) Protein- and calcium-mediated GLP-1 secretion: a narrative review. Adv Nutr 12: 2540–2552. https://doi.org/10.1093/advances/nmab078
11. Watkins JD et al (2022) Plasma glucagon-like peptide-1 responses to ingestion of protein with increasing doses of milk minerals rich in calcium. Br J Nutr 128:55–63. https://doi.org/10.1017/s000711452100297x
12. Bode HP et al (1999) Glucagon-like peptide 1 elevates cytosolic calcium in pancreatic beta-cells independently of protein kinase A. Endocrinology 140:3919–3927. https://doi.org/10.1210/endo.140.9.6947
13. Akimbekov NS et al (2024) The role of magnesium in pancreatic beta-cell function and homeostasis. Front Nutr 11:1458700. https://doi.org/10.3389/fnut.2024.1458700
14. AlAnazi MM et al (2024) A systematic review of the gene-lifestyle interactions on metabolic disease-related outcomes in Arab populations. Nutrients 16:2519. https://doi.org/10.3390/nu16152519
15. Ahmad R et al (2024) Zinc and diabetes: a connection between micronutrient and metabolism. Cells 13:1359. https://doi.org/10.3390/cells13161359
16. Mironov N et al (2022) Phosphate dysregulation and metabolic syndrome. Nutrients 14:4477. https://doi.org/10.3390/nu14214477
17. Buckeridge C et al (2024) Once-daily oral small-molecule glucagon-like peptide-1 receptor agonist lotiglipron (PF-07081532) for type 2 diabetes and obesity: two randomized, placebo-controlled, multiple-ascending-dose Phase 1 studies. Diabetes Obes Metab 26:3155–3166. https://doi.org/10.1111/dom.15643
18. Shaefer CF Jr et al (2015) User's guide to mechanism of action and clinical use of GLP-1 receptor agonists. Postgrad Med 127:818–826. https://doi.org/10.1080/00325481.2015.1090295
19. Hinnen D (2017) Glucagon-like peptide 1 receptor agonists for type 2 diabetes. Diabetes Spectr 30:202–210. https://doi.org/10.2337/ds16-0026
20. Hosseinpour A et al (2024) Glucagon-like peptide-1 receptor agonists and major adverse cardiovascular events in patients with and without diabetes: a meta-analysis of randomized-controlled trials. Clin Cardiol 47:e24314. https://doi.org/10.1002/clc.24314
21. Sodhi M et al (2023) Risk of gastrointestinal adverse events associated with glucagon-like peptide-1 receptor agonists for weight loss. JAMA 330:1795–1797. https://doi.org/10.1001/jama.2023.19574
22. Guyton J et al (2019) Glucagon-like peptide 1 receptor agonists in type 1 diabetes mellitus. Am J Health Syst Pharm 76:1739–1748. https://doi.org/10.1093/ajhp/zxz179
23. Helmstadter J et al (2022) Glucagon-like peptide-1 (GLP-1) receptor agonists and their cardiovascular benefits-the role of the GLP-1 receptor. Br J Pharmacol 179:659–676. https://doi.org/10.1111/bph.15462
24. Rowlands J et al (2018) Pleiotropic effects of GLP-1 and analogs on cell signaling, metabolism, and function. Front Endocrinol (Lausanne) 9:672. https://doi.org/10.3389/fendo.2018.00672
25. Standaert DG (2024) GLP-1, Parkinson's disease, and neuroprotection. N Engl J Med 390:1233–1234. https://doi.org/10.1056/NEJMe2401743
26. Meissner WG et al (2024) Trial of lixisenatide in early Parkinson's disease. N Engl J Med 390:1176–1185. https://doi.org/10.1056/NEJMoa2312323
27. Veniant MM et al (2024) A GIPR antagonist conjugated to GLP-1 analogues promotes weight loss with improved metabolic parameters in preclinical and phase 1 settings. Nat Metab 6:290–303. https://doi.org/10.1038/s42255-023-00966-w
28. Petersen J et al (2024) GLP-1-directed NMDA receptor antagonism for obesity treatment. Nature 629: 1133–1141. https://doi.org/10.1038/s41586-024-07419-8
29. Jin FX et al (2024) Intestinal glucagon-like peptide-1: a new player associated with impaired counterregulatory responses to hypoglycaemia in type 1 diabetic mice. World J Diabetes 15:1764–1777. https://doi.org/10.4239/wjd.v15.i8.1764
30. Liu LR et al (2025) Intestinal glucagon-like peptide-1: a new regulator of impaired counterregulatory responses to hypoglycemia in type 1 diabetes mellitus. World J Diabetes 16:99726. https://doi.org/10.4239/wjd.v16.i1.99726
31. Snell-Bergeon JK, Wadwa RP (2012) Hypoglycemia, diabetes, and cardiovascular disease. Diabetes Technol Ther 14(Suppl 1):S51–S58. https://doi.org/10.1089/dia.2012.0031
32. Almby KE et al (2019) Effects of GLP-1 on counter-regulatory responses during hypoglycemia after GBP surgery. Eur J Endocrinol 181:161–171. https://doi.org/10.1530/eje-19-0171
33. Lim GE et al (2009) Insulin regulates glucagon-like peptide-1 secretion from the enteroendocrine L cell. Endocrinology 150:580–591. https://doi.org/10.1210/en.2008-0726
34. Mozaffarian D (2024) GLP-1 agonists for obesity—a new recipe for success? JAMA 331:1007–1008. https://doi.org/10.1001/jama.2024.2252
35. Drucker DJ (2024) The benefits of GLP-1 drugs beyond obesity. Science 385:258–260. https://doi.org/10.1126/science.adn4128

Maternal and Fetal Magnesium Balance: Impacts and Implications

Desiree Lanehart, Chantal Uzoma, Md S. Zaman, Nuraly S. Akimbekov, Juan C. Lopez-Alvarenga, and Mohammed S. Razzaque

Abstract

Magnesium (Mg^{2+}) is an essential mineral that supports normal cellular metabolism, optimal musculoskeletal and neuronal functions, the immune response, and cardiovascular health. The regulation of magnesium is a complex process, and during pregnancy, it becomes even more challenging because of increased magnesium needs. Magnesium inadequacy during pregnancy can lead to severe complications, including preeclampsia in the mother, intrauterine growth restriction, preterm birth, low birth weight, insulin resistance, heart disease, and developmental delays in the neonate, due in part to its critical role in fetal musculoskeletal and nervous development. Magnesium deficiency during pregnancy is a global health concern, and raising awareness about the importance of maintaining magnesium balance, improving dietary intake, and promoting supplementation, when needed, could help reduce associated risks. This chapter discusses the impact of altered magnesium homeostasis on maternal health and fetal development while also addressing the need for global awareness and interventions to improve maternal and child health.

Keywords

Magnesium · Pregnancy · Maternal health · Fetal development · Child health

Magnesium Homeostasis

Magnesium is an important mineral in the human body, and it is crucial in many essential processes. Studies indicate that magnesium metabolism involves more than 600 different biochemical reactions [1, 2] related to neurogenesis, the production of DNA and RNA, energy production, protein synthesis, muscle and bone development, and cardiovascular and immune functions [1, 3, 4]. Despite extensive studies, the key

D. Lanehart · C. Uzoma · M. S. Razzaque (✉)
Department of Medical Education, School of Medicine, University of Texas Rio Grande Valley (UTRGV), Edinburg, TX, USA
e-mail: mohammed.razzaque@utrgv.edu

M. S. Zaman
Department of Biological Sciences, Alcorn State University, Lorman, MS, USA

Department of Biology, South Texas College, McAllen, TX, USA

N. S. Akimbekov
International Center for Islamic Science and Innovation, Al-Farabi Kazakh National University, Almaty, Kazakhstan

Scientific-Practical Center, West Kazakhstan Marat Ospanov Medical University, Aktobe, Kazakhstan

J. C. Lopez-Alvarenga
Division of Population Health & Biostatistics, School of Medicine, University of Texas Rio Grande Valley, Edinburg, TX, USA

M. S. Razzaque (ed.), *Vitamin D Function*, Advances in Experimental Medicine and Biology 1493,
https://doi.org/10.1007/978-3-032-04357-3_11

contributions of magnesium to health and well-being remain unexplored.

The recommended dietary allowance (RDA) of magnesium for pregnant women is between 350 and 400 mg/day, which is approximately 10% higher than that of nonpregnant or nonlactating women [2]. A woman's body weight affects the optimal concentration of the supplement. Those with a higher BMI may have lower magnesium levels after receiving standard doses, indicating that additional dosing or administration methods might be needed to maintain efficient levels [5]. The balance between magnesium intake, excretion, and cellular distribution is important. As mentioned earlier, magnesium plays a key role in many important functions, including maintaining muscle and nerve functions, heart rate, blood pressure, energy production, metabolic reactions, immune regulation, nucleic acid synthesis, and protein synthesis [6, 7]. Approximately 60% of the body's total magnesium is found in bones, 25% is found in muscles, and the remainder is found in soft tissues, tissue fluids, and the bloodstream. Blood contains only approximately 1% magnesium, which makes regulating magnesium crucially important to ensure normal body function [8, 9]. Magnesium from food is absorbed in the intestines, and its levels in the blood are controlled mainly by the kidneys. When magnesium intake is low, the amount of magnesium excreted in the urine is reduced to conserve it. When magnesium intake is high, the kidneys increase magnesium excretion to prevent excess buildup in the body [10]. This tight control ensures that magnesium is available in sufficient quantities for essential functions while preventing imbalances. Magnesium deficiency is also a commonly encountered global health concern. Many people suffer from low magnesium levels, especially mothers and newborns living in lower-income regions with limited access to adequate nutrition, increased stressors, and a lack of healthcare infrastructure [11]. This highlights the need for public health initiatives to address magnesium deficiency, especially in pregnant women. As previously highlighted, the implications of hypomagnesemia are vast and detrimental to both mothers and infants. According to the World Health Organization (WHO), magnesium deficiency affects a large portion of the global population, and pregnant women in these regions are especially vulnerable. Low magnesium intake, poor dietary patterns, and inadequate healthcare infrastructure contribute to maternal magnesium deficiency, with serious implications for maternal and fetal health [12].

Magnesium deficiency during pregnancy remains a concern even in developed and high-income countries, especially in individuals with underlying health conditions such as diabetes, hypertension, or poor nutritional status [13]. As mentioned earlier, magnesium deficiency has also been linked to an increased risk of preeclampsia, which is a leading cause of maternal morbidity and mortality globally [14]. The relationship between magnesium levels and pregnancy-related complications highlights the need for routine screening and intervention to prevent hypomagnesemia in pregnant women. Public health strategies must move beyond awareness campaigns to include targeted interventions such as prenatal screening for magnesium deficiency, community nutrition programs, and food fortification in low-resource populations. Implementing scalable supplementation initiatives along with these other targeted interventions could meaningfully reduce rates of preeclampsia, low birth weight, and preterm birth. This chapter explores the balance of magnesium in the body during pregnancy, the transfer of magnesium from the mother to the fetus, and the effects of magnesium deficiency on both maternal and fetal health.

Importance of Magnesium Transfer from Mother to Fetus During Pregnancy

Understanding how magnesium is balanced in the body and transferred from the mother to the fetus is essential for recognizing the effects of magnesium on both the mother and the developing newborn. Improving uteroplacental blood flow allows magnesium to reduce the risk of intrauterine growth restriction (IUGR) and preterm labor. Additionally, its anti-inflammatory properties have been linked to a lower incidence of necrotizing enterocolitis (NEC), a serious disease causing inflamed intestinal lining in premature infants [15].

Magnesium is mostly stored in bones and tissues, with only a small amount circulating in the bloodstream [16, 17]. The body delicately controls magnesium levels by balancing how much is taken in, the amount excreted, and its distribution to cells. This balance becomes more complicated during pregnancy, as the mother's body needs more magnesium to support the growing fetus [2]. Magnesium is transferred against a concentration gradient using active transport. Magnesium sulfate crosses the placenta easily, achieving a 1:1 ratio between maternal and fetal magnesium levels. However, magnesium levels are usually higher in the fetus than in the mother. This is especially true during the third trimester, when placental transfer increases to the fetus to aid in the maturation of the fetal skeleton [5]. This efficient transfer is vital for the maintenance of adequate levels of magnesium in the fetus, for without it, the risk of neonatal complications rises markedly. Notably, when a substance crosses the placenta to enter the fetal circulation, its concentration or amount in the maternal circulation may decrease. This highlights the importance of monitoring both maternal and fetal health during magnesium sulfate therapy. Research has also shown a decline in magnesium levels in mothers from preconception to the third trimester, some of which is due to increased magnesium transport from the mother to the fetus (an average of 4.5 mg/day) [18]. Even within the current literature, the intricacies of maternal-fetal magnesium transfer and the balance between too much and too little supplementation are not well understood [2]. While this process is delicately regulated, maternal magnesium deficiency can impair this transfer leading to fetal hypomagnesaemia characterized by magnesium serum concentration lower than 1.46 mg/dL [2, 18]. Beyond its role in growth and development, magnesium also plays a role in antioxidant defense—an area critically relevant to conditions such as necrotizing enterocolitis (NEC). NEC is a major surgical emergency in the neonatal intensive care unit (NICU), affecting 1–3 newborns per 1000 live births. A study revealed that magnesium and copper deficiencies impact the antioxidant defense of the gut and increase the production of pro-NEC cytokines [19]. Normally, the fetal antioxidant system increases late in pregnancy to prepare for oxidative challenges after birth [20]. In premature infants, this process is shortened, and a lack of antioxidants such as magnesium worsens the defense. This connection between oxidative stress, nutrient deficiency, and prematurity undermines the potential for preventive strategies to reduce the incidence of NEC [20].

Magnesium on Fetal Neurodevelopment and Impacts of Hypomagnesemia

Adequate magnesium is crucial for facilitating proper neuronal differentiation and synaptic plasticity and reducing neuronal excitotoxicity. Magnesium ensures proper embryonic development by providing neuroprotection [21]. Reflecting its broader neurological benefits, studies have reported reduced incidence and mortality associated with pediatric cerebral palsy [22]. Magnesium has been shown to prevent astrocyte death and reduce glial fibrillary acidic protein (GFAP) secretion in preterm deliveries. Magnesium successfully lowers GFAP levels, which supports a protective effect on the neonatal brain (Fig. 1) [23].

Ion channels, such as TRPM6 and TRPM7 (genes that are involved in the production of channels that transfer magnesium), are vital for ion exchange during embryonic development [2]. Without these channels, severe developmental defects or fetal death can occur. However, magnesium can bypass the need for these channels by passive diffusion through paracellular transport and protect alternative pathways. It also aids in DNA replication and cellular division, which are essential for organ development.

One of the most significant impacts of magnesium deficiency on the fetus is the risk of intrauterine growth restriction (IUGR), a condition

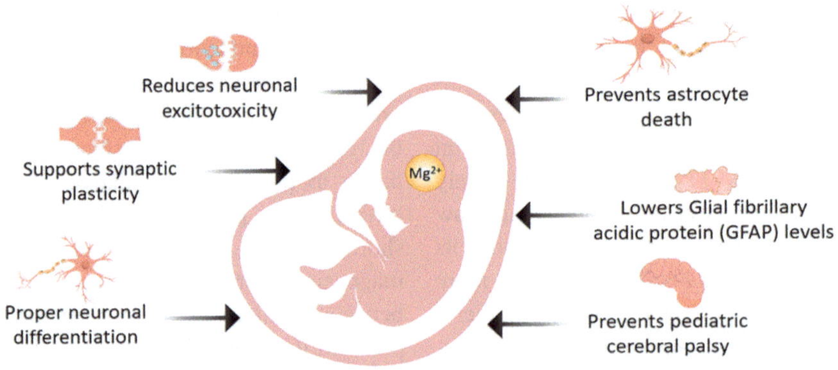

Fig. 1 Role of magnesium in providing neuroprotection during embryonic development

where a fetus does not grow as expected for its gestational age. Newborns with IUGR are at greater risk of feeding difficulties, respiratory distress syndrome, cardiovascular issues, motor skills and coordination, and impaired brain development [24]. Magnesium is also important for maintaining healthy electrolyte balance, supporting muscle function, and ensuring proper neuromuscular control [25]. Infants born with low magnesium levels may experience symptoms such as tremors, irritability, and seizures [26]. These immediate challenges can have a significant effect on an infant's health in the early stages of life. Furthermore, magnesium plays a role in regulating calcium and potassium levels, both of which are important for proper muscle function [1]. Magnesium deficiency can result in muscle weakness or abnormal muscle development in the fetus, which may lead to problems with movement, coordination, and physical development after birth [2]. In extreme cases, magnesium deficiency during pregnancy may increase the risk of skeletal deformities or bone abnormalities in the child [13].

Magnesium also regulates the release of neurotransmitters for proper neuronal signaling [27]. Low magnesium levels during pregnancy have been linked to an increased risk of neurological problems in the child, such as cognitive delays, learning disabilities, and motor deficits [28]. Magnesium deficiency may also impair the formation of synapses, impairing communication throughout the nervous system. The long-term effects of impaired brain development can extend into childhood and beyond, potentially leading to lifelong learning challenges and behavioral issues [28, 29].

Studies have also shown that hypomagnesemia is involved in long-term sequelae suggesting an increased risk of developing chronic diseases later in life [2]. Magnesium plays a key role in regulating insulin sensitivity. As a result, individuals who are exposed to magnesium deficiency in utero may be at increased risk of developing conditions such as type 2 diabetes and osteoporosis as they grow older [30]. Epidemiological data link fetal hypomagnesemia to a greater risk of insulin resistance later in life, likely due to IUGR. However, prevention methods are still unclear. A "small-for-gestational-age (SGA) and preterm children" study focusing on insulin and glucose regulation revealed an increased risk of hyperinsulinemia after glucose loads [31]. These findings suggest that hypomagnesemia might indirectly contribute to metabolic issues, particularly in vulnerable populations.

Pharmacological Agents Influencing Magnesium Deficiency

While magnesium deficiency is often attributed to insufficient intake, many external factors, such as drug interactions, increased excretion of magnesium by diuretics, underlying gastrointestinal problems, and increasing physiological demands during gestation, can exacerbate this deficiency as well. Certain medications, such as proton pump inhibitors (PPIs), reduce stomach acid, decreasing the absorptive capacity of magnesium [32]. In these instances, it is important to either supplement with additional magnesium or titrate the conflicting medications to lower dosages.

PPIs inhibit magnesium absorption independently in combination with diuretics [2]. Others have shown that PPIs may not affect passive absorption, but they do reduce the efficacy of the active transport system [32]. Malabsorption conditions and certain dietary habits, such as chronic alcoholism or ketogenic diets, can impair the uptake of magnesium by intestinal enterocytes. Contributing factors such as altered gut permeability, a reduction in transporter proteins, and competition with other ions can interfere with magnesium absorption to the point of severely decreased levels if not detected early [33].

Hypomagnesemia can seriously affect the fetus, but magnesium levels must be corrected carefully. Without proper monitoring, especially in patients with kidney issues or obesity, intravenous magnesium sulfate can lead to dangerously high magnesium levels in mothers. Low magnesium levels are linked to high BMI, which can worsen both metabolic problems and magnesium deficiency. In some cases, increasing the dose for a higher BMI may cause toxicity, particularly in people with kidney issues. Hypermagnesemia can cause symptoms in mothers, such as nausea, vomiting, and lethargy, along with more serious issues, such as bradycardia, hypotension, and even cardiac arrest, in severe cases. Excess magnesium affects calcium flow in the heart, leading to slower signal transmission and reduced heart function [34–36]. If untreated, fetal magnesium toxicity can lead to bone issues and weak muscle tone. A University of Texas Southwestern Medical Center study revealed that high magnesium levels were linked to weak reflexes, respiratory issues, lower Apgar scores, and more NICU admissions. It can also cause feeding intolerance and gastrointestinal problems, impairing nutrition, weakening the immune system, and increasing the risk of severe complications [37].

Clinical Manifestations and Management of Magnesium Imbalance

While manifestations of magnesium imbalance can vary widely, they often include neurologic and systemic symptoms such as tremors, seizures (refractory to calcium or anticonvulsants), hypocalcemia [secondary to impaired parathyroid hormone (PTH) release], poor feeding or vomiting, apathy, and hypotonia. Laboratory values are usually <1.5 mg/dL, often with low concentrations of calcium and potassium and metabolic alkalosis (in renal wasting syndromes). Of relevance, magnesium is required for normal PTH secretion. Parathyroid hormone is responsible for the regulation of calcium and phosphorus levels in the blood by altering absorption of these nutrients in the kidneys, bones, and intestines. If magnesium concentration is severely low (i.e., <0.8 mg/dL), the parathyroid glands stop secreting PTH. In acute cases, intravenous magnesium sulfate is needed to combat this issue.

During pregnancy, hemodilution of magnesium levels may occur. For supplementation, magnesium oxide (400–800 mg/day of elemental magnesium) is commonly used, although magnesium citrate or lactate is better tolerated. In acute cases, intravenous magnesium sulfate (4–6 g over 20–30 min) is recommended, followed by a continuous infusion of 1–2 g/h. For preeclampsia and eclampsia, various regimens exist, including IV, intramuscular (IM), or combined IV/IM protocols. In cases of magnesium toxicity—indicated by a respiratory rate <12 breaths/min, absence of deep tendon reflexes, urine output <30 mL/h, or serum magnesium >8 mg/dL—administer 10 mL of 10% calcium gluconate IV over 10 min.

Conclusions

Magnesium is essential for maternal health and fetal development during pregnancy [38]. Optimal magnesium balance is required for fetal

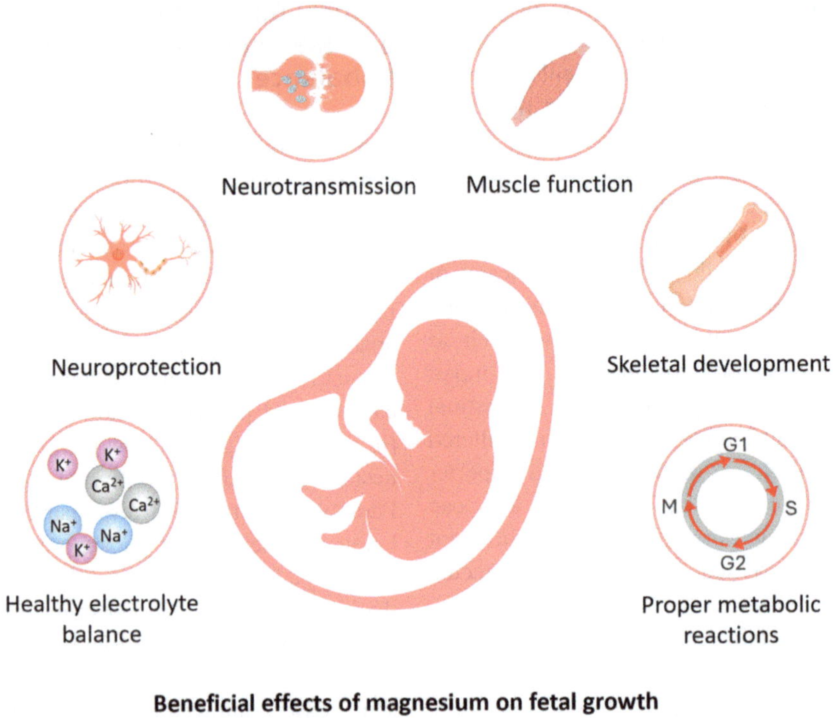

Fig. 2 Beneficial effects of magnesium on fetal growth and development

musculoskeletal growth, neuronal development, and maintaining metabolic and electrolyte homeostasis (Fig. 2). Deficiencies can lead to complications for both mothers and infants, particularly in low-income regions. Even in more developed countries, some women remain at risk. Public health programs focused on nutrition and magnesium supplementation could help reduce complications and improve health outcomes for both mothers and infants. Future research should focus on improving magnesium intake, particularly for vulnerable populations, and develop policies to monitor and address magnesium deficiency in pregnant women globally. Effective public health interventions enhance prenatal care, and targeted programs in underserved regions are crucial for improving maternal and fetal health outcomes.

Acknowledgments D.L. and C.U. are Doctor of Medicine (MD) students at the School of Medicine, University of Texas Rio Grande Valley (UTRGV), Edinburg, Texas, USA. Information has been collected from online sources, including OpenAI GPT-3.5 and Google Scholar. The authors are solely responsible for the content. M.S.R. is a visiting professor at the University Kuala Lumpur Royal College of Medicine Perak, Ipoh, Malaysia.

References

1. Al Alawi AM, Majoni SW, Falhammar H (2018) Magnesium and human health: perspectives and research directions. Int J Endocrinol 2018:9041694
2. Fanni D, Gerosa C, Nurchi VM, Manchia M, Saba L, Coghe F, Crisponi G, Gibo Y, Van Eyken P, Fanos V, Faa G (2021) The role of magnesium in pregnancy and in fetal programming of adult diseases. Biol Trace Elem Res 199:3647–3657
3. Fatima G, Dzupina A, Alhmadi HB, Magomedova A, Siddiqui Z, Mehdi A, Hadi N (2024) Magnesium matters: a comprehensive review of its vital role in health and diseases. Cureus 16:e71392
4. Liao W, Jiang M, Li M, Jin C, Xiao S, Fan S, Fang W, Zheng Y, Liu J (2017) Magnesium elevation promotes neuronal differentiation while suppressing glial differentiation of primary cultured adult mouse neural progenitor cells through ERK/CREB activation. Front Neurosci 11:87
5. Brookfield KF, Su F, Elkomy MH, Drover DR, Lyell DJ, Carvalho B (2016) Pharmacokinetics and placental transfer of magnesium sulfate in pregnant women. Am J Obstet Gynecol 214(737):e731–e739

6. Eskander M, Razzaque MS (2022) Can maintaining optimal magnesium balance reduce the disease severity of COVID-19 patients? Front Endocrinol (Lausanne) 13:843152
7. Uwitonze AM, Rahman S, Ojeh N, Grant WB, Kaur H, Haq A, Razzaque MS (2020) Oral manifestations of magnesium and vitamin D inadequacy. J Steroid Biochem Mol Biol 200:105636
8. Erem S, Atfi A, Razzaque MS (2019) Anabolic effects of vitamin D and magnesium in aging bone. J Steroid Biochem Mol Biol 193:105400
9. Razzaque MS (2018) Magnesium: are we consuming enough? Nutrients 10:1863
10. Jahnen-Dechent W, Ketteler M (2012) Magnesium basics. Clin Kidney J 5:i3–i14
11. Ladipo OA (2000) Nutrition in pregnancy: mineral and vitamin supplements. Am J Clin Nutr 72:280S–290S
12. Shukla V, Parvez S, Fatima G, Singh S, Magomedova A, Batiha GE, Alexiou A, Papadakis M, Welson NN, Hadi N (2024) Micronutrient interactions: magnesium and its synergies in maternal-fetal health. Food Sci Nutr 12:6913–6928
13. Naowar M, Dickton D, Francis J (2024) Cardiometabolic risk factors associated with magnesium and vitamin D nutrients during pregnancy-a narrative review. Nutrients 16:2630
14. Duley L, Gulmezoglu AM, Chou D (2010) Magnesium sulphate versus lytic cocktail for eclampsia. Cochrane Database Syst Rev 2010:CD002960
15. Roman A, Desai N, Rochelson B, Gupta M, Solanki M, Xue X, Chatterjee PK, Metz CN (2013) Maternal magnesium supplementation reduces intrauterine growth restriction and suppresses inflammation in a rat model. Am J Obstet Gynecol 208(383):e381–e387
16. Akimbekov NS, Coban SO, Atfi A, Razzaque MS (2024) The role of magnesium in pancreatic beta-cell function and homeostasis. Front Nutr 11:1458700
17. Uwitonze AM, Razzaque MS (2018) Role of magnesium in vitamin D activation and function. J Am Osteopath Assoc 118:181–189
18. Morton A (2018) Hypomagnesaemia and pregnancy. Obstet Med 11:67–72
19. Maayan-Metzger A, Itzchak A, Mazkereth R, Kuint J (2004) Necrotizing enterocolitis in full-term infants: case-control study and review of the literature. J Perinatol 24:494–499
20. Aceti A, Beghetti I, Martini S, Faldella G, Corvaglia L (2018) Oxidative stress and necrotizing enterocolitis: pathogenetic mechanisms, opportunities for intervention, and role of human milk. Oxid Med Cell Longev 2018:7397659
21. Patel V, Akimbekov NS, Grant WB, Dean C, Fang X, Razzaque MS (2024) Neuroprotective effects of magnesium: implications for neuroinflammation and cognitive decline. Front Endocrinol (Lausanne) 15: 1406455
22. Crowther CA, Hiller JE, Doyle LW, Haslam RR, Australasian Collaborative Trial of Magnesium Sulphate Collaborative Group (2003) Effect of magnesium sulfate given for neuroprotection before preterm birth: a randomized controlled trial. JAMA 290:2669–2676
23. Santoso DPJ, Nugrahani AD, Siddiq A, Pramatirta AY, Aziz MA, Irianti S, Pribadi A, Anwar AD, Effendi JS (2023) Effect of maternal serum magnesium and calcium levels on umbilical glial fibrillary acidic protein levels in preterm labor. Sci Rep 13: 13337
24. Pike K, Jane Pillow J, Lucas JS (2012) Long term respiratory consequences of intrauterine growth restriction. Semin Fetal Neonatal Med 17:92–98
25. Souza ACR, Vasconcelos AR, Dias DD, Komoni G, Name JJ (2023) The integral role of magnesium in muscle integrity and aging: a comprehensive review. Nutrients 15:5127
26. Srinivasan R, Ninama AC, Krishna S, Vasudevan A (2023) Familial hypomagnesemia with hypocalcemia: a rare cause of infantile seizures. CEN Case Rep 12: 139–145
27. Kirkland AE, Sarlo GL, Holton KF (2018) The role of magnesium in neurological disorders. Nutrients 10: 730
28. Cortes-Albornoz MC, Garcia-Guaqueta DP, Velez-van-Meerbeke A, Talero-Gutierrez C (2021) Maternal nutrition and neurodevelopment: a scoping review. Nutrients 13:3530
29. Xue W, You J, Su Y, Wang Q (2019) The effect of magnesium deficiency on neurological disorders: a narrative review article. Iran J Public Health 48:379–387
30. Barbagallo M, Dominguez LJ (2015) Magnesium and type 2 diabetes. World J Diabetes 6:1152–1157
31. Chan PY, Morris JM, Leslie GI, Kelly PJ, Gallery ED (2010) The long-term effects of prematurity and intrauterine growth restriction on cardiovascular, renal, and metabolic function. Int J Pediatr 2010:280402
32. Florentin M, Elisaf MS (2012) Proton pump inhibitor-induced hypomagnesemia: a new challenge. World J Nephrol 1:151–154
33. Swaminathan R (2003) Magnesium metabolism and its disorders. Clin Biochem Rev 24:47–66
34. Aal-Hamad AH, Al-Alawi AM, Kashoub MS, Falhammar H (2023) Hypermagnesemia in clinical practice. Medicina (Kaunas) 59:1190
35. Cascella M, Vaqar S (2025) Hypermagnesemia. In: StatPearls. StatPearls Publishing, Treasure Island
36. Ghaddar M, Hatab T, El-Kaakour A, Tamim H, Makki M, El-Halabi T, Rifaii K, Sfeir P, Hoteit M, Koubar SH (2025) Risk of postoperative hypermagnesemia in cardiopulmonary bypass-assisted cardiovascular surgery. Kidney 360(6):219–226
37. Belden MK, Gnadt S, Ebert A (2017) Effects of maternal magnesium sulfate treatment on neonatal feeding tolerance. J Pediatr Pharmacol Ther 22:112–117
38. Chiarello DI, Marín R, Proverbio F, Coronado P, Toledo F, Salsoso R, Gutiérrez J, Sobrevia L (2018) Mechanisms of the effect of magnesium salts in preeclampsia. Placenta 69:134–139

Vitamin D-Induced Antimicrobial Peptides in Combating Viral Infections

Nuraly S. Akimbekov, Ilya Digel, Kuanysh Tastambek, Rea Rodriguez-Raecke, Aida S. Kistaubayeva, Svetlana K. Sakhanova, Xia Wu, and Mohammed S. Razzaque

Abstract

Viral infectious diseases rank among the leading causes of mortality globally. The rise of drug resistance mechanisms puts the effectiveness of all existing antimicrobial drugs at risk, several of which have become ineffective. Consequently, there is an acute need for novel antimicrobial agents. Host defense antimicrobial peptides (AMPs) are naturally occurring, evolutionarily conserved peptides that contribute to innate immunity and exhibit broad activity against viruses, bacteria, and fungi. They can also exert immunomodulatory and adjuvant activities by serving as chemotactic agents for immune cells and promoting cytokine and chemokine production. Vitamin D is a fat-soluble steroid hormone essential for regulating calcium and phosphorus metabolism to preserve bone homeostasis. Recent research has indicated that vitamin D has immunomodulatory effects on both the innate and adaptive immune systems, suggesting that it is a vital regulator of immunological homeostasis. Vitamin D stimulates the development of robust AMPs found in natural killer cells, monocytes, neutrophils, and epithelial cells of

N. S. Akimbekov (✉)
International Center for Islamic Science and Innovation, Al-Farabi Kazakh National University, Almaty, Kazakhstan

Scientific-Practical Center, West Kazakhstan Marat Ospanov Medical University, Aktobe, Kazakhstan

Ecology Research Institute, Khoja Akhmet Yassawi International Kazakh-Turkish University, Turkistan, Kazakhstan
e-mail: Akimbekov.Nuraly@kaznu.edu.kz

I. Digel
Institute for Bioengineering, Aachen University of Applied Sciences, Jülich, Germany

K. Tastambek
International Center for Islamic Science and Innovation, Al-Farabi Kazakh National University, Almaty, Kazakhstan

Ecology Research Institute, Khoja Akhmet Yassawi International Kazakh-Turkish University, Turkistan, Kazakhstan

R. Rodriguez-Raecke
Interdisciplinary Center for Clinical Research (IZKF), RWTH Aachen University, Aachen, Germany

A. S. Kistaubayeva
International Center for Islamic Science and Innovation, Al-Farabi Kazakh National University, Almaty, Kazakhstan

S. K. Sakhanova
Scientific-Practical Center, West Kazakhstan Marat Ospanov Medical University, Aktobe, Kazakhstan

X. Wu
School of Food and Biological Engineering, Shaanxi University of Science and Technology, Xi'an, Shaanxi, China

M. S. Razzaque
Department of Medical Education, School of Medicine, University of Texas Rio Grande Valley (UTRGV), Edinburg, TX, USA

M. S. Razzaque (ed.), *Vitamin D Function*, Advances in Experimental Medicine and Biology 1493,
https://doi.org/10.1007/978-3-032-04357-3_12

the respiratory tract. This chapter explains the evidence suggesting that vitamin D can reduce viral survival and replication by inducing AMPs.

Keywords

Antimicrobial peptide · Cathelicidin · Defensins · Viral infection · Innate immunity · Vitamin D · Vitamin D receptor

Introduction

In recent years, the world has experienced several major viral infectious diseases, including severe acute respiratory syndrome coronavirus 2 (SARS-CoV-2), other respiratory infections, dengue fever, rabies, and acute hepatitis. Although these diseases arise from different viral families, they share a common characteristic: their ability to cause complications not only through direct damage to target tissues but also by triggering an excessive inflammatory response by the human host to infection [1].

It has long been suggested that vitamin D deficiency may increase susceptibility to viral infections, as vitamin D levels are correlated with the severity of the immune response to these infections [2]. Epidemiological studies conducted over the past two decades have demonstrated a strong correlation between seasonal variations in vitamin D levels and a higher incidence of viral infectious diseases [3, 4]. Research has revealed that various immune system cells express the vitamin D receptor (VDR) and respond to the action of 1,25 $(OH)_2D$, confirming that it is a significant regulator of both innate and adaptive immune responses [5, 6]. Studies have shown that the vitamin D-VDR complex regulates the expression of genes involved in the production of AMPs, such as cathelicidins, defensins, and hepcidins [1, 7]. Cathelicidin LL-37 is a particularly potent AMP that is lethal to a broad spectrum of pathogens, including various viruses (Fig. 1).

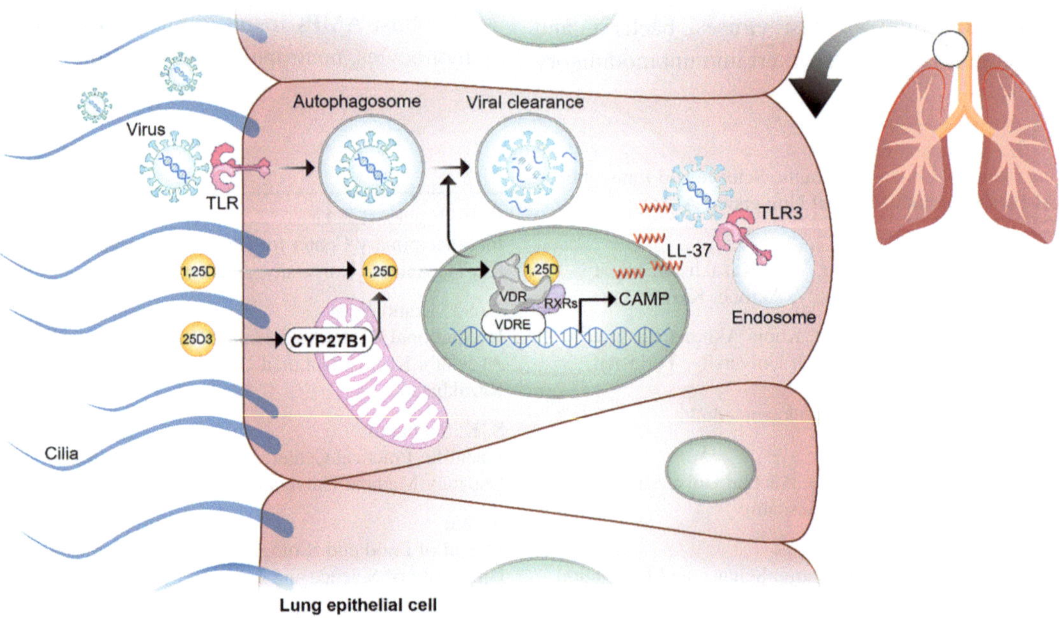

Fig. 1 Antiviral properties of vitamin D-induced AMPs in lung epithelial cells. The figure depicts the processes involved in vitamin D metabolism and signaling during a viral infection, as well as the responses triggered by the active form of vitamin D, $1,25(OH)_2D$. One key response is the induction of cathelicidin antimicrobial peptide (CAMP) gene expression. Its mature product, LL-37, binds to the viral genome, facilitating efficient binding to endosomal toll-like receptor (TLR3), which enhances TLR3 signaling and promotes viral clearance. Additionally, $1,25(OH)_2D$ induces autophagy, providing another mechanism for the removal of viruses

Vitamin D also plays an active role in the differentiation of antigen-presenting cells, influences the development of dendritic cells with tolerogenic properties, and affects antigen presentation by these cells [8]. In the context of the adaptive immune system, vitamin D is recognized as a key participant in the activation of cell-mediated (Th1) and humoral (Th2) immunity [9]. It regulates the expression of both pro-inflammatory and anti-inflammatory interleukins and inhibits the differentiation and proliferation of antibody-producing B cells [10]. Several studies have proposed a link between a lack of vitamin D and the occurrence of specific infections because of its impact on the immune system [11, 12]. They also proposed vitamin D treatment as a preventive measure for complications related to these infections. Vitamin D deficiency has been linked to adverse outcomes in individuals with HIV, higher rates of influenza A infections, and increased respiratory infections [13–15]. Supplementation with vitamin D has been found to reduce the occurrence of seasonal flu among school children, elderly individuals, and African-American women and to decrease the severity of respiratory tract infections [16–18].

AMPs and Their Antiviral Activity

AMPs provide protection from viral infection both directly by acting on the virion and indirectly by interfering with viral proliferation and replication in the target cell. Furthermore, defensin expression can be induced by cytokines, which can modulate adaptive immune responses [19–21]. Human defensins are known to be effective against both enveloped and nonenveloped viruses (Table 1). The activity of LL-37 against some viruses has also been demonstrated despite the commonly held perception that it is effective only against bacterial infections.

The therapeutic potential of peptide-based antiviral drugs is becoming increasingly evident. It has already been argued to be efficient, cost-effective, and safe because it prompts peptide elimination following treatment and has a low incidence of side effects [70]. In fact, some AMPs have already been approved by the US Food and Drug Administration (FDA) as antiviral drugs against HIV-1, HSV, VZV, CMV, and IAV [71].

Antiviral Mechanisms of AMPs

Research has demonstrated that the antiviral mechanisms exhibited by AMPs are highly diverse and impact all stages of the viral life cycle [19] (Fig. 2). Interactions between AMPs and viruses primarily occur in two main ways: first, on the virus envelope, where AMP integration can lead to membrane instability, thereby hindering the virus's ability to infect host cells [72], and second, with viral glycoproteins, which prevents the virus from interacting with its target cells [73].

The first point of interference for cationic and amphipathic AMPs is the negatively charged lipid bilayer of the enveloped viruses. Not all enveloped viruses are affected by these AMPs, and the range of viruses they can target can be highly unpredictable. Only the α-defensin HNP1 has been shown to directly act against HSV-1, and only the β-defensin hBD2 has been proven to act directly against RSV, deactivating viruses by targeting their envelope [65]. Defensins HNP1–3, HD5, and LL-37 display lectin-like qualities, which allow them to interfere with the viral glycoproteins that facilitate the binding and fusion of viruses to host cells [19]. HSV infection is prevented by the action of the α-, β-, and θ-defensins HNP1–4, HD5, HD6, hBD3, and LL-37, primarily through the inhibition of viral attachment and entry processes targeting HSV glycoprotein B (gB) and heparan sulfate [45].

AMPs can also prevent viral spread indirectly by interacting with host cells in addition to directly targeting the virus. Viral entry into host cells can be blocked in two ways: first, by competing with the virus for receptor binding to the host cells, thus preventing infection [74], and second, by entering the host cells and interacting with cytoplasmic organelles, leading to changes in gene expression and inhibiting cell-to-cell

Table 1 Antiviral effects of AMPs

Virus			AAV	CMV	HAdV[a]	HIV-1	HRV	HPV	HSV-1/-2	IAV	PyV[a]	RSV	VSV	VV	VZV
LL-37						+ [22, 23]				+ [24, 25]		+ [26, 27]		+ [28–30]	+ [31]
Defensins	HNP	1	+ [32]	+ [33]	+ [34–38]	+ [39–42]		+ [43]	+ [33, 44, 45]	+ [33, 46–48]	+ [49]		+ [33]		
		2	+ [32]			+ [50]		+ [43]	+ [33, 44, 45]	+ [46, 47, 51]					
		3				+ [50]		+ [43]	+ [33, 44, 45]						
		4				+ [52]			+ [45]						
	HD	5	+ [53]		+ [34–36, 38, 54, 55]			+ [43, 54]	+ [45]	+ [47]	+ [49]				
		6	+ [53]						+ [45]						
	HBD	1		+ [56]						+ [57]					
		2			+ [36, 38]	+ [58–61]	+ [62, 63]	+ [64]				+ [65]			+ [31]
		3				+ [58–61, 66]	+ [62, 63, 67]		+ [45]	+ [68]				+ [69]	
		4													

AAV adeno-associated virus, *CMV* cytomegalovirus, *HAdV* human adenovirus, *HIV* human immunodeficiency virus, *HRV* human rhinovirus, *HPV* human papillomavirus, *HSV* herpes simplex virus, *IAV* influenza A virus, *PyV* polyomavirus, *RSV* respiratory syncytial virus, *VSV* vesicular stomatitis virus, *VV* vaccinia virus, *VZV* varicella zoster virus

[a]Depending on serotypes

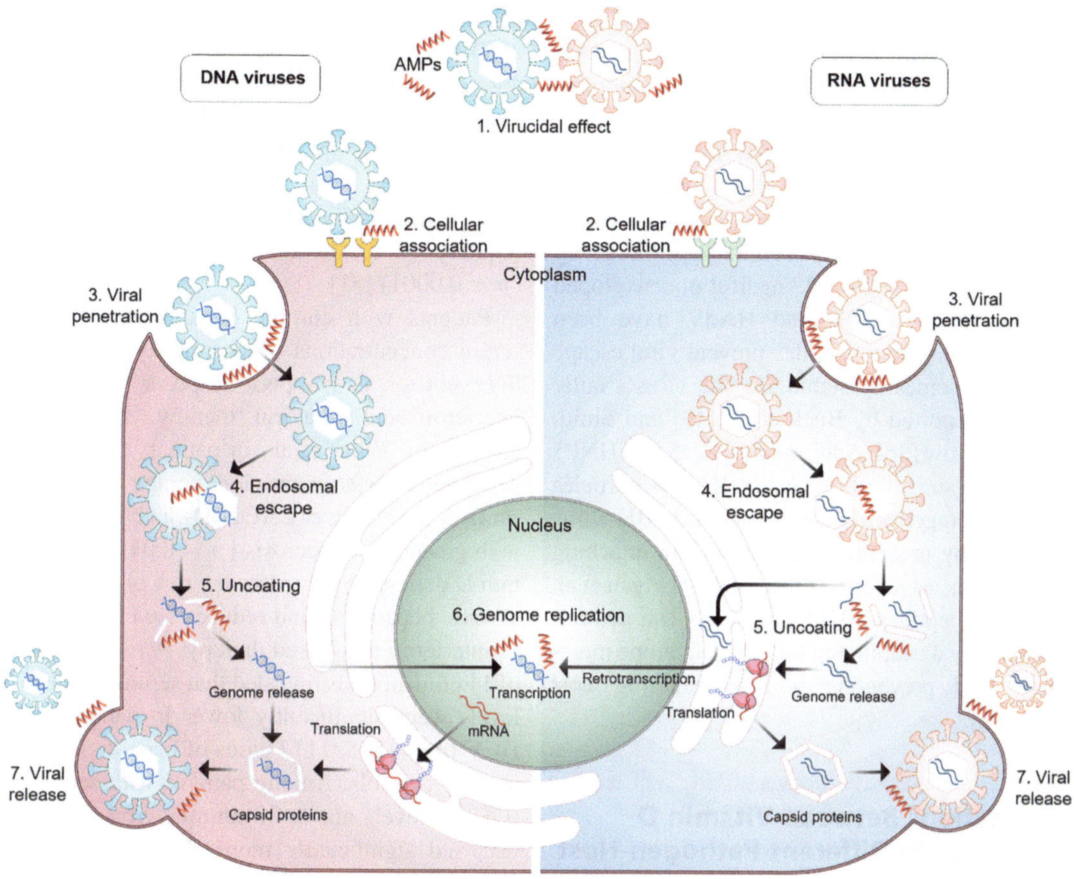

Fig. 2 AMPs play crucial roles in inhibiting various stages of the DNA and RNA viral replication cycle. AMPs that have a described mechanism of action can be categorized according to their inhibition sites as follows: (1) Virucidal AMPs act directly to inhibit viral particles. (2) By interfering with viral interactions, AMPs disrupt viral interactions and subsequent adsorption. (3) With respect to viral penetration, AMPs prevent the penetration of viruses into host cells. (4) AMPs inhibit the escape of the virus from the endosome. (5) AMPs affect the uncoating process of the virus. (6) AMPs suppress viral gene expression. (7) AMPs inhibit the release of mature virions

transmission [75, 76]. Defensins HNP1–4, hBD2–3, and LL-37 have been shown to neutralize HIV-1 infection by interfering with the virus's life cycle at various points. Notably, LL-37 has been demonstrated to block HIV-1 replication by suppressing reverse transcriptase activity [22]. HIV-1 infection is also inhibited by hBD2 and hBD3 in three key ways: direct inactivation of the virus, modulation of the CXCR4 coreceptor, and suppression of viral replication [50, 59].

In contrast, HNP α-defensins, particularly HNP1, can exert different effects throughout the HIV-1 virus life cycle to prevent infection. The binding ability of the CD4 receptor and the HIV envelope protein gp120 to their molecules has been reported [41] to cause the downregulation of CD4 and CXCR4 coreceptors. Alternative mechanisms of action proposed for HNP1 include prevention of viral internalization, suppression of gene expression, and inhibition of reverse transcription and integration—partly through interference with protein kinase C [45, 77]. Studies by Wilson et al. and Tripathi et al. have shown that IAV is vulnerable to the antiviral effects of certain defensins, including the α-, β-, and θ-defensins

HNP1–4, HD5–6, hBD1–3, and LL-37 [19, 25]. Specifically, LL-37 may prevent IAV infection by damaging the viral membrane, whereas α-defensins either inhibit the function of protein kinase C, an enzyme involved in IAV escape and entry into the nucleus and viral trafficking, or cause viral particles to clump together [71]. Mechanisms of action exhibited by HNP1–4, HD5, and LL-37 against nonenveloped viruses such as HPV and HAdV have been documented. These peptides prevent viral escape from endosomes by stabilizing the virus's outer shell, as reported by Buck et al. [43] and Smith and Nemerow [36]. Additionally, HD5 and HNP1 have been suggested to cause BK virus particles to clump together, thereby blocking BK virus infection by preventing the virus from attaching to host cells, as reported in studies by Dugan et al. [49]. The peptides hBD2 and LL-37 can neutralize RSV by destabilizing the viral envelope membrane and preventing it from entering host cells [26].

Interactions Between Vitamin D and AMPs in Different Pathogen-Host Systems

In the following sections, the reported data concerning the effects of vitamin D against various viral infections are briefly addressed, with a particular emphasis on its association with AMPs.

Vitamin D, AMPs, and Viral Hepatitis

Vitamin D is known to suppress hepatitis C virus (HCV) replication [78, 79]. In a study by Huang et al., HCV-infected HuH-7 cells were cultured with and without vitamin D. The results demonstrated that 25(OH)D_3 reduced the extra- and intracellular levels of the main HCV antigen [80]. The inhibitory effect of 25(OH)D_3 occurs during the emergence of the infecting viral complex; additionally, 25(OH)D_3 suppresses HCV replication and promotes a rapid virological response in the body. Cell survival and the viral load were observed in the Con1-1b and J6/JFH-2a cell lines in culture treated with different doses of 25(OH)D_3. The viral load of Con1-1b cells was dose dependent, with a 69–86% reduction after treatment with 1, 5, or 10 μM 25(OH)D ($p < 0.0001$). In cells, similar results were obtained: The viral load decreased by 12%, 55%, and 80.5% after 1, 5, and 10 μM 25(OH)D_3 treatment, respectively ($p < 0.0001$) [81].

Patients with chronic hepatitis C with low serum concentrations of vitamin D have severe liver fibrosis and a poor response to standard interferon and ribavirin therapy. In a meta-analysis by Villar et al., in patients treated with pegylated interferon antiviral therapy with ribavirin ($n = 167$), lower 25(OH)D_3 levels correlated with greater liver necrosis ($p = 0.04$) [82]. Vitamin D deficiency increases the risk of developing chronic hepatitis C and reduces the effectiveness of interferon alfa-based therapy. A meta-analysis of 11 clinical trials revealed that serum 25(OH)D_3 levels were significantly lower in patients with HCV ($n = 1575$, 1117 cases of hepatitis) than in healthy controls. Among patients with serum 25(OH)D_3 levels above 30 ng/mL, viral clearance occurred significantly more often (hazard ratio (HR) 1.57, 95% confidence interval (CI) 1.12–2.2) [83].

A meta-analysis of seven randomized controlled trials ($n = 548$) revealed that adjuvant antiviral therapy for HCV is more effective when vitamin D is added. Compared with therapy without vitamin D, the use of vitamin D in combination with interferon alfa injections and oral ribavirin for 24 weeks resulted in a 30% improvement in treatment efficacy. This effect was particularly notable in patients with HCV genotype 1, which is typically resistant to antiviral therapy [84].

Vitamin D, AMPs, and Chronic Lung Diseases

For preschool children, Jensen et al. [85] demonstrated an inverse relationship between the severity of virus-induced asthma attacks, the frequency of exacerbation, the dose of inhaled corticosteroids (ICS) on one side, and the serum $25(OH)D_3$ level on the other side. Other randomized trials also confirmed that daily administration of preventive doses of vitamin D or large doses once per month helps reduce the incidence of lung and asthma attacks and, correspondingly, the dose of ICS [86, 87]. Aregbesola et al. [88] examined the risk factors for hospitalization for pneumonia in an aging population in eastern Finland. The investigators reported an inverse correlation between serum $25(OH)D_3$ levels and the risk of pneumonia. The reports published by Mamani et al. [89] and Brance et al. [90] revealed an inverse correlation between $25(OH)D_3$ and the severity of community-acquired pneumonia, as measured by the CURB-65 index (mental confusion, uremia, respiratory rate, blood pressure, age ≥65 years). The most obvious mechanism that closely links vitamin D and the frequency and/or severity of chronic lung disease exacerbation is vitamin D's ability to induce the expression of AMPs, especially HBD-2 and LL-37 [91, 92].

Vitamin D and AMPs Against Seasonal Influenza

A few studies suggest a role for vitamin D supplementation in preventing influenza A virus (IAV) infection. Jones et al. [93] examined 46 blood samples from adults and children, some of whom were experiencing influenza virus infections. The levels of retinol-binding protein (RBP), vitamin D, and other antibodies were measured. In 44 patients, RBP and vitamin D deficiency or insufficiency were detected. In addition, vitamin D levels were correlated with blood IgM and IgG3, whereas RBP levels were correlated with IgG4 and IgA.

LL-37 is secreted by respiratory epithelial cells on the airway surface to form a key line of defense against foreign pathogens [94]. The antiviral activity of LL-37 has also been reported against both enveloped and nonenveloped viruses, including IAV and respiratory syncytial virus (RSV) [71]. Barlow et al. [24] reported that LL-37 has antiviral activities against IAV both in vitro and in vivo. Although the exact mechanisms by which cathelicidins modulate inflammation during IAV infection are not clear, the modulation of the TLR-mediated inflammatory response by LL-37 has been well demonstrated [95]. α-Defensins have also been shown to neutralize and aggregate IAV and promote the uptake of the virus by neutrophils [51].

Vitamin D, AMPs, and Respiratory Syncytial Virus

Respiratory syncytial virus (RSV) infection weakens the body's antiviral defense through its effects on vitamin D metabolism [96, 97]. RSV infection reduces the expression of the vitamin D receptor and increases the expression of the enzyme 1-beta-hydroxylase, which inactivates vitamin D [98]. Vitamin D promotes antiviral protection of bronchial epithelial cells through the action of the antiviral and antibacterial peptide cathelicidin and activation of interferon signaling pathways [99]. By supporting antiviral immunity, vitamin D also reduces RSV-induced activation of the pro-inflammatory factor NF-kB and associated cytokines in the airway epithelium. It induces the protein IkBa, an inhibitor of the NF-kB factor, and suppresses the RSV-induced activation of pro-inflammatory genes controlled by NF-kB (interferon beta, CXCL10, etc.). The inhibition of NF-kB by adenoviral particles containing IkBa mimicked the effects of vitamin D. Thus, vitamin D attenuated the inflammatory response to viral infections without compromising antiviral activity [100]. A meta-analysis of 25 randomized trials ($n = 10{,}933$) using individual patient data revealed that vitamin D supplementation resulted in an average 12% reduction in the risk of acute respiratory infection (OR 0.88, 95% CI 0.81–0.96, $p < 0.001$). Furthermore, subgroup analysis revealed that among patients who had more than one episode of acute respiratory illness per year, the risk of recurrent illness was reduced by 20% with vitamin D supplementation (OR 0.80, 95% 0.69–0.93; $p = 0.004$) [101].

Vitamin D, AMPs, and Acute Respiratory Infections

Vitamin D deficiency can be considered a "seasonal stimulus" for the development of acute respiratory infections (ARIs) [15, 102, 103]. Given that influenza epidemics at midlatitudes are most intense during winter, it is reasonable to assume that seasonal fluctuations in vitamin D levels affect the human immune system. Vitamin D levels are typically low during the influenza season and highest in summer, when the incidence of ARI is minimal [104]. One of the most notable findings supporting the protective role of vitamin D in ARIs is the stimulation of AMP production by 1,25$(OH)_2D$. These endogenous peptides act directly to destroy respiratory viruses. Other mechanisms by which vitamin D reduces the risk of ARIs include the inhibition of pro-inflammatory Th1 cytokine production (e.g., tumor necrosis factor-α and interferon-γ) and the promotion of anti-inflammatory cytokine expression by macrophages [9, 105].

Lang and Samaras [106] reported that a serum 25$(OH)D_3$ level of 30 ng/mL is necessary for optimal induction of cathelicidin mRNA, although levels above 40 ng/mL did not produce additional effects.

The antiviral activity of vitamin D is also linked to its ability to modulate the inflammatory response. The synthesis of pro-inflammatory cytokines in response to influenza virus infection appears to be correlated with the severity of the disease [102]. In a study by Khare et al. [107] using human lung A549 epithelial cells, a significant decrease in the production of TNF-α, IFN-β, IL-6, and IL-8 was observed during infection with influenza A (H1N1) virus following the administration of 1,25$(OH)_2D$. Helming et al. [108] revealed the ability of 1,25$(OH)_2D$ to reduce the production of pro-inflammatory cytokines by modulating macrophage activity, preventing the excessive secretion of various cytokines and chemokines. The authors proposed that 1,25 $(OH)_2D$ is involved in negative feedback regulation, in which interferon γ-activated macrophages induce the release of 1,25$(OH)_2D$. Once a sufficient concentration is reached, the expression of VDR and its translocation into the nucleus are synergistically upregulated. Moreover, it can suppress important IFN-γ-induced genes, such as Ccl5, Cxcl10, Cxcl9, Irf2, Fcgr1, Fcgr3, and Tlr2. Moan et al. [109] compared the seasonality of deaths from influenza and pneumonia in Norway with that of serum vitamin D levels. In the course of the study, the authors concluded that increased mortality from influenza and pneumonia in winter is associated with low levels of vitamin D, supporting the hypothesis that vitamin D acts as a factor in the prevention and control of ARIs. In a double-blind, placebo-controlled trial, Laaksi et al. [110] studied 164 healthy young men (aged 18–28 years) undergoing military training. The participants were randomized to receive either 400 IU of vitamin D daily ($n = 80$) or placebo ($n = 84$). After 6 months, the mean serum 25$(OH)D_3$ concentration in the intervention group was 71.6 ± 22.9 nmol/L ($n = 58$), whereas it was 51.3 ± 15.5 nmol/L in the placebo group ($n = 50$) ($p < 0.001$). The primary endpoint was the number of days absent from duty due to an ARI. The proportion of participants without ARIs during the study period was significantly greater in the vitamin D group (51.3%) than in the placebo group (35.5%) ($p = 0.045$). These results confirmed the preventive effect of vitamin D supplementation against respiratory tract infections. The authors also noted that the risk ratio for work disability due to respiratory infection was lower in the vitamin D group.

A clinical study including 18,883 patients revealed a pronounced inverse association between the frequency of ARIs and serum 25 $(OH)D_3$ levels [111]. This effect may be especially relevant in patients with underlying respiratory conditions such as asthma and chronic obstructive pulmonary disease. An extensive epidemiological study by Urashima et al. examined the relationship between vitamin D deficiency and the development of diseases associated with type A influenza virus [16]. Compared with control children, schoolchildren receiving vitamin D supplements were two times less likely to catch the flu—10.8% and 18.6%, respectively ($p = 0.04$). The beneficial role of vitamin D in ARIs and its effect on lung function have been confirmed by Berry et al. [112], who used cross-sectional data from 6789 participants in the nationwide 1958 British birth cohort. The investigators measured serum levels of 25 $(OH)D_3$, lung function (forced expiratory volume and forced vital capacity), and the frequency of ARI from the age of 45 years and revealed a linear relationship between vitamin D deficiency, seasonal infections, and lung function. An increase in the concentration of $25(OH)D_3$ for each 10 nmol/L increase was associated with a 7% reduction in the risk of developing ARI.

Vitamins A and D are both known to support a healthy mucosal immune response in mice, particularly through their role in stimulating IgA, a critical component of mucosal defense. In a case-control study by Nanri et al. [113] among employees of four Japanese companies, unvaccinated participants with adequate vitamin D levels ($\geq$30 ng/mL $25(OH)D_3$) had a lower risk of influenza infection in the winter. Among vaccinated individuals, no significant correlation was found between serum $25(OH)D_3$ levels and influenza incidence.

The Role of Vitamin D and AMPs in the Treatment of COVID-19

The concept that vitamin D directly influences viral infections, especially those produced by enveloped viruses such as SARS-CoV-2, is supported by numerous cell culture studies [114, 115]. The antiviral function of vitamin D is associated with its upregulatory effects on cathelicidins, which are released by neutrophils, macrophages, and lung epithelial cells (such as LL-37), as well as β-defensin-2 (BD-2), which may be induced by reactive oxygen species [116, 117]. Cathelicidins regulate both innate and adaptive immunity, primarily their effects on monocytes, dendritic cells, and T cells [118]. The capacity of LL-37 to suppress viruses mostly stems from its direct impact on the viral envelope [25, 119]. LL-37 likely inhibits viral entry into host cells in a manner analogous to other antimicrobial peptides. LL-37 has been shown to exhibit antiviral activity against HSV-1, VACV, retroviruses, RSV, and certain nonenveloped viruses [120]. These findings support the concept that the antiviral properties of LL-37 are at least partly mediated through envelope disruption [24, 121]. Given that SARS-CoV-2 is an enveloped virus, modulating circulating vitamin D levels may aid in COVID-19 treatment by increasing LL-37 production in the respiratory tract. LL-37, which is detected in bronchoalveolar lavage fluid, may support viral envelope disruption and influence apoptotic and immunological pathways [118].

Niv [122] reported that human defensin 5 (HD5) has a much greater affinity for the mucosal cell membrane protein angiotensin-converting enzyme 2 (ACE2) (39.3 nM) than the SARS-CoV-2 spike protein (14.7 nM). In addition, HD-5 may bind to the glycosylated SARS-CoV-2 S1 protein, potentially enhancing its protective efficacy. Although pretreatment with HD-5 has been demonstrated to decrease SARS-CoV-2 pseudoparticle titers, its effect on SARS-CoV-2 isolates has not yet been established [123]. A review by Grand et al. [121] suggested that vitamin D may reduce the risk of COVID-19 infection through several mechanisms: maintaining epithelial barrier integrity via tight junctions, inducing innate immunity through AMP production, and inhibiting pro-inflammatory cytokine production, thereby reducing the risk of cytokine storm and related pneumonia.

The antiviral properties of LL-37 using SARS-CoV-2 pseudovirions have also been explored. Wang et al. [124] reported that LL-37 can bind to the spike protein and cloak the cellular SARS-CoV-2 receptor ACE2. Another study demonstrated [125] that LL-37 was an effective inhibitor in mouse models infected with spike protein-expressing SARS-CoV-2 pseudovirions. In the same study, HBD-2 was also shown to selectively block infection by these pseudovirions. Since vitamin D induces the production of both cathelicidins and defensins [91, 126], it may serve as a promising therapeutic agent in antiviral strategies targeting COVID-19.

Conclusion

Despite substantial scientific evidence supporting the role of vitamin D in activating and regulating immune responses, as well as its potential protective effects against viral infections and related inflammatory complications, significant knowledge gaps still exist that need to be addressed. Future research should aim to clarify the specific mechanisms through which vitamin D modulates immune functions. This knowledge could support the development of novel treatment and prevention strategies based on antimicrobial peptides for infectious diseases.

References

1. Galindo-Méndez M, Galindo-Ruiz M, Concheso-Venegas MF, Mendoza-Molina SU, Orozco-Cruz D, Weintraub-Benzion E (2025) The impact of vitamin D in the prevention of influenza, COVID-19, and dengue: a review. Biomedicine 13:927
2. Siddiqui M, Manansala JS, Abdulrahman HA, Nasrallah GK, Smatti MK, Younes N, Althani AA, Yassine HM (2020) Immune modulatory effects of vitamin D on viral infections. Nutrients 12:2879
3. Danai PA, Sinha S, Moss M, Haber MJ, Martin GS (2007) Seasonal variation in the epidemiology of sepsis*. Crit Care Med 35:410
4. Grant WB (2008) Variations in vitamin D production could possibly explain the seasonality of childhood respiratory infections in Hawaii. Pediatr Infect Dis J 27:853
5. Penna G, Amuchastegui S, Giarratana N, Daniel KC, Vulcano M, Sozzani S, Adorini L (2007) 1,25-Dihydroxyvitamin D3 selectively modulates tolerogenic properties in myeloid but not plasmacytoid dendritic cells. J Immunol 178:145–153
6. Mahon BD, Wittke A, Weaver V, Cantorna MT (2003) The targets of vitamin D depend on the differentiation and activation status of CD4 positive T cells. J Cell Biochem 89:922–932
7. Shin D-M, Jo E-K (2011) Antimicrobial peptides in innate immunity against mycobacteria. Immune Netw 11:245–252
8. Piemonti L, Monti P, Sironi M, Fraticelli P, Leone BE, Dal Cin E, Allavena P, Di Carlo V (2000) Vitamin D3 affects differentiation, maturation, and function of human monocyte-derived dendritic cells. J Immunol 164:4443–4451
9. Gombart AF, Pierre A, Maggini S (2020) A review of micronutrients and the immune system–working in harmony to reduce the risk of infection. Nutrients 12:236
10. Hossein-Nezhad A, Mirzaei K, Keshavarz SA, Ansar H, Saboori S, Tootee A (2013) Evidences of dual role of vitamin D through cellular energy homeostasis and inflammation pathway in risk of cancer in obese subjects. Minerva Med 104:295–307
11. Pitman MC, Meagher N, Price DJ, Rhodes A, Chang JJ, Scher B, Allan B, Street A, McMahon JH, Rasmussen TA, Cameron PU, Hoy JF, Kent SJ, Lewin SR (2023) Effect of high dose vitamin D3 on the HIV-1 reservoir: a pilot randomised controlled trial. J Virus Erad 9:100345
12. Papagni R, Pellegrino C, Di Gennaro F, Patti G, Ricciardi A, Novara R, Cotugno S, Musso M, Guido G, Ronga L, Stolfa S, Bavaro DF, Romanelli F, Totaro V, Lattanzio R, De Iaco G, Palmieri F, Saracino A, Gualano G (2022) Impact of vitamin D in prophylaxis and treatment in tuberculosis patients. Int J Mol Sci 23:3860
13. Bodnar LM, Krohn MA, Simhan HN (2009) Maternal vitamin D deficiency is associated with bacterial vaginosis in the first trimester of pregnancy. J Nutr 139:1157–1161
14. Villamor E (2006) A potential role for vitamin D on HIV infection? Nutr Rev 64:226–233
15. Sabetta JR, DePetrillo P, Cipriani RJ, Smardin J, Burns LA, Landry ML (2010) Serum 25-hydroxyvitamin d and the incidence of acute viral respiratory tract infections in healthy adults. PLoS One 5:e11088
16. Urashima M, Segawa T, Okazaki M, Kurihara M, Wada Y, Ida H (2010) Randomized trial of vitamin D supplementation to prevent seasonal influenza A in schoolchildren. Am J Clin Nutr 91:1255–1260
17. Avenell A, Cook JA, Maclennan GS, Macpherson GC (2007) Vitamin D supplementation to prevent infections: a sub-study of a randomised placebo-controlled trial in older people (RECORD trial, ISRCTN 51647438). Age Ageing 36:574–577
18. Kenny EE, Pe'er I, Karban A, Ozelius L, Mitchell AA, Ng SM, Erazo M, Ostrer H, Abraham C, Abreu MT, Atzmon G, Barzilai N, Brant SR, Bressman S, Burns ER, Chowers Y, Clark LN, Darvasi A, Doheny D, Duerr RH, Eliakim R, Giladi N, Gregersen PK, Hakonarson H, Jones MR, Marder K, McGovern DP, Mulle J, Orr-Urtreger A, Proctor DD, Pulver A, Rotter JI, Silverberg MS, Ullman T, Warren ST, Waterman M, Zhang W, Bergman A, Mayer L, Katz S, Desnick RJ, Cho JH, Peter I (2012) A genome-wide scan of Ashkenazi Jewish Crohn's disease suggests novel susceptibility loci. PLoS Genet 8:e1002559
19. Wilson SS, Wiens ME, Smith JG (2013) Antiviral mechanisms of human defensins. J Mol Biol 425:4965–4980
20. Park MS, Kim JI, Lee I, Park S, Bae J-Y, Park M-S (2018) Towards the application of human defensins as antivirals. Biomol Ther 26:242
21. Hancock REW, Sahl H-G (2006) Antimicrobial and host-defense peptides as new anti-infective therapeutic strategies. Nat Biotechnol 24:1551–1557
22. Bergman P, Walter-Jallow L, Broliden K, Agerberth B, Soderlund J (2007) The antimicrobial peptide LL-37 inhibits HIV-1 replication. Curr HIV Res 5:410–415

23. Levinson P, Kaul R, Kimani J, Ngugi E, Moses S, MacDonald KS, Broliden K, Hirbod T, Kibera HIVSG (2009) Levels of innate immune factors in genital fluids: association of alpha defensins and LL-37 with genital infections and increased HIV acquisition. AIDS 23:309–317
24. Barlow PG, Svoboda P, Mackellar A, Nash AA, York IA, Pohl J, Davidson DJ, Donis RO (2011) Antiviral activity and increased host defense against influenza infection elicited by the human cathelicidin LL-37. PLoS One 6:e25333
25. Tripathi S, Tecle T, Verma A, Crouch E, White M, Hartshorn KL (2013) The human cathelicidin LL-37 inhibits influenza A viruses through a mechanism distinct from that of surfactant protein D or defensins. J Gen Virol 94:40–49
26. Currie SM, Findlay EG, McHugh BJ, Mackellar A, Man T, Macmillan D, Wang H, Fitch PM, Schwarze J, Davidson DJ (2013) The human cathelicidin LL-37 has antiviral activity against respiratory syncytial virus. PLoS One 8:e73659
27. Mansbach JM, Piedra PA, Borregaard N, Martineau AR, Neuman MI, Espinola JA, Camargo CA (2012) Serum cathelicidin level is associated with viral etiology and severity of bronchiolitis. J Allergy Clin Immunol 130:1007–1008
28. Howell MD, Gallo RL, Boguniewicz M, Jones JF, Wong C, Streib JE, Leung DYM (2006) Cytokine milieu of atopic dermatitis skin subverts the innate immune response to vaccinia virus. Immunity 24: 341–348
29. Howell MD, Jones JF, Kisich KO, Streib JE, Gallo RL, Leung DYM (2004) Selective killing of vaccinia virus by LL-37: implications for eczema vaccinatum. J Immunol 172:1763–1767
30. Dean RE, O'Brien LM, Thwaite JE, Fox MA, Atkins H, Ulaeto DO (2010) A carpet-based mechanism for direct antimicrobial peptide activity against vaccinia virus membranes. Peptides 31:1966–1972
31. Crack LR, Jones L, Malavige GN, Patel V, Ogg GS (2012) Human antimicrobial peptides LL-37 and human β-defensin-2 reduce viral replication in keratinocytes infected with varicella zoster virus. Clin Exp Dermatol 37:534–543
32. Virella-Lowell I, Poirier A, Chesnut KA, Brantly M, Flotte TR (2000) Inhibition of recombinant adeno-associated virus (rAAV) transduction by bronchial secretions from cystic fibrosis patients. Gene Ther 7:1783–1789
33. Daher KA, Selsted ME, Lehrer RI (1986) Direct inactivation of viruses by human granulocyte defensins. J Virol 60:1068–1074
34. Smith JG, Silvestry M, Lindert S, Lu W, Nemerow GR, Stewart PL (2010) Insight into the mechanisms of adenovirus capsid disassembly from studies of defensin neutralization. PLoS Pathog 6:e1000959
35. Lu W, de Leeuw E (2013) Pro-inflammatory and pro-apoptotic properties of human defensin 5. Biochem Biophys Res Commun 436:557–562
36. Smith JG, Nemerow GR (2008) Mechanism of adenovirus neutralization by human α-defensins. Cell Host Microbe 3:11–19
37. Bastian A, Schäfer H (2001) Human α-defensin 1 (HNP-1) inhibits adenoviral infection in vitro. Regul Pept 101:157–161
38. Gropp R, Frye M, Wagner TOF, Bargon J (1999) Epithelial defensins impair adenoviral infection: implication for adenovirus-mediated gene therapy. Hum Gene Ther 10:957–964
39. Wei G, Pazgier M, de Leeuw E, Rajabi M, Li J, Zou G, Jung G, Yuan W, Lu W-Y, Lehrer RI (2010) Trp-26 imparts functional versatility to human α-defensin HNP1. J Biol Chem 285:16275–16285
40. Chang TL, Vargas J, DelPortillo A, Klotman ME (2005) Dual role of α-defensin-1 in anti–HIV-1 innate immunity. J Clin Invest 115:765–773
41. Demirkhanyan LH, Marin M, Padilla-Parra S, Zhan C, Miyauchi K, Jean-Baptiste M, Novitskiy G, Lu W, Melikyan GB (2012) Multifaceted mechanisms of HIV-1 entry inhibition by human α-defensin*♦. J Biol Chem 287:28821–28838
42. Zhang J, Liu Z, Zhou Z, Huang Z, Yang Y, Wu J, Liu Y (2025) HNP-1: from structure to application thanks to multifaceted functions. Microorganisms 13:458
43. Buck CB, Day PM, Thompson CD, Lubkowski J, Lu W, Lowy DR, Schiller JT (2006) Human α-defensins block papillomavirus infection. Proc Natl Acad Sci 103:1516–1521
44. Ganz T, Selsted ME, Szklarek D, Harwig SS, Daher K, Bainton DF, Lehrer RI (1985) Defensins. Natural peptide antibiotics of human neutrophils. J Clin Invest 76:1427–1435
45. Hazrati E, Galen B, Lu W, Wang W, Ouyang Y, Keller MJ, Lehrer RI, Herold BC (2006) Human α-and β-defensins block multiple steps in herpes simplex virus infection. J Immunol 177:8658–8666
46. Hartshorn KL, White MR, Tecle T, Holmskov U, Crouch EC (2006) Innate defense against influenza A virus: activity of human neutrophil defensins and interactions of defensins with surfactant protein D. J Immunol 176:6962–6972
47. Tecle T, White MR, Gantz D, Crouch EC, Hartshorn KL (2007) Human neutrophil defensins increase neutrophil uptake of influenza A virus and bacteria and modify virus-induced respiratory burst responses. J Immunol 178:8046–8052
48. Salvatore M, García-Sastre A, Ruchala P, Lehrer RI, Chang T, Klotman ME (2007) β-Defensin inhibits influenza virus replication by cell-mediated mechanism(s). J Infect Dis 196:835–843
49. Dugan AS, Maginnis MS, Jordan JA, Gasparovic ML, Manley K, Page R, Williams G, Porter E, O'Hara BA, Atwood WJ (2008) Human α-defensins inhibit BK virus infection by aggregating virions and blocking binding to host cells. J Biol Chem 283: 31125–31132

50. Weinberg A, Quinones-Mateu ME, Lederman MM (2006) Role of human β-defensins in HIV infection. Adv Dent Res 19:42–48
51. Doss M, White MR, Tecle T, Gantz D, Crouch EC, Jung G, Ruchala P, Waring AJ, Lehrer RI, Hartshorn KL (2009) Interactions of alpha-, beta-, and theta-defensins with influenza A virus and surfactant protein D. J Immunol 182:7878–7887
52. Wu Z, Cocchi F, Gentles D, Ericksen B, Lubkowski J, DeVico A, Lehrer RI, Lu W (2005) Human neutrophil α-defensin 4 inhibits HIV-1 infection in vitro. FEBS Lett 579:162–166
53. Zhao C, Porter JM, Burke PC, Arnberg N, Smith JG (2024) Alpha-defensin binding expands human adenovirus tropism. PLoS Pathog 20:e1012317
54. Gounder AP, Wiens ME, Wilson SS, Lu W, Smith JG (2012) Critical determinants of human α-defensin 5 activity against non-enveloped viruses. J Biol Chem 287:24554–24562
55. Nguyen EK, Nemerow GR, Smith JG (2010) Direct evidence from single-cell analysis that human α-defensins block adenovirus uncoating to neutralize infection. J Virol 84:4041–4049
56. Duan J, Wang H, Liu M, Chen Y, Li N, Liu J, Wang L, Li L, Liu Y, Dong P, Wang X, Fan Z, Jiao S (2024) Tumor-derived DEFB1 induces immune tolerance by inhibiting maturation of dendritic cell and impairing CD8+ T cell function in esophageal squamous cell carcinoma. Chin J Cancer Res 36:351–367
57. Ryan LK, Dai J, Yin Z, Megjugorac N, Uhlhorn V, Yim S, Schwartz KD, Abrahams JM, Diamond G, Fitzgerald-Bocarsly P (2011) Modulation of human β-defensin-1 (hBD-1) in plasmacytoid dendritic cells (PDC), monocytes, and epithelial cells by influenza virus, Herpes simplex virus, and Sendai virus and its possible role in innate immunity. J Leukoc Biol 90: 343–356
58. Herrera R, Morris M, Rosbe K, Feng Z, Weinberg A, Tugizov S (2016) Human beta-defensins 2 and-3 cointernalize with human immunodeficiency virus via heparan sulfate proteoglycans and reduce infectivity of intracellular virions in tonsil epithelial cells. Virology 487:172–187
59. Quiñones-Mateu ME, Lederman MM, Feng Z, Chakraborty B, Weber J, Rangel HR, Marotta ML, Mirza M, Jiang B, Kiser P (2003) Human epithelial β-defensins 2 and 3 inhibit HIV-1 replication. AIDS 17:F39–F48
60. Sun L, Finnegan CM, Kish-Catalone T, Blumenthal R, Garzino-Demo P, La Terra Maggiore GM, Berrone S, Kleinman C, Wu Z, Abdelwahab S (2005) Human β-defensins suppress human immunodeficiency virus infection: potential role in mucosal protection. J Virol 79:14318–14329
61. Zapata W, Rodriguez B, Weber J, Estrada H, Quiñones-Mateu ME, Zimermman PA, Lederman MM, Rugeles MT (2008) Increased levels of human beta-defensins mRNA in sexually HIV-1 exposed but uninfected individuals. Curr HIV Res 6:531–538
62. Duits LA, Nibbering PH, van Strijen E, Vos JB, Mannesse-Lazeroms SPG, van Sterkenburg MAJA, Hiemstra PS (2003) Rhinovirus increases human β-defensin-2 and-3 mRNA expression in cultured bronchial epithelial cells. FEMS Immunol Med Microbiol 38:59–64
63. Proud D, Sanders SP, Wiehler S (2004) Human rhinovirus infection induces airway epithelial cell production of human β-defensin 2 both in vitro and in vivo. J Immunol 172:4637–4645
64. Chen H, Guo K, Bai Z, Lu L, Liu B, Zhang J, Zhong M, Xu C, Chen W, Huang A, Ding Y (2025) Advances in the prevention of cervical cancer by anti-human papillomavirus agents. Cancer Med 14: e70847
65. Kota S, Sabbah A, Harnack R, Xiang Y, Meng X, Bose S (2008) Role of human β-defensin-2 during tumor necrosis factor-α/NF-κB-mediated innate antiviral response against human respiratory syncytial virus. J Biol Chem 283:22417–22429
66. Feng Z, Dubyak GR, Lederman MM, Weinberg A (2006) Cutting edge: human β defensin 3—a novel antagonist of the HIV-1 coreceptor CXCR4. J Immunol 177:782–786
67. Mamun TI, Ali MA, Hosen MN, Rahman J, Islam MA, Akib MG, Zaman K, Rahman MM, Hossain FMA, Ibenmoussa S, Bourhia M, Dawoud TM (2025) Designing a multi-epitope vaccine candidate against human rhinovirus C utilizing immunoinformatics approach. Front Immunol 15: 1364129
68. Leikina E, Delanoe-Ayari H, Melikov K, Cho M-S, Chen A, Waring AJ, Wang W, Xie Y, Loo JA, Lehrer RI (2005) Carbohydrate-binding molecules inhibit viral fusion and entry by crosslinking membrane glycoproteins. Nat Immunol 6:995–1001
69. Howell MD, Streib JE, Leung DYM (2007) Antiviral activity of human β-defensin 3 against vaccinia virus. J Allergy Clin Immunol 119:1022–1025
70. Thakur N, Qureshi A, Kumar M (2012) AVPpred: collection and prediction of highly effective antiviral peptides. Nucleic Acids Res 40:W199–W204
71. Gwyer Findlay E, Currie SM, Davidson DJ (2013) Cationic host defence peptides: potential as antiviral therapeutics. BioDrugs 27:479–493
72. Sitaram N, Nagaraj R (1999) Interaction of antimicrobial peptides with biological and model membranes: structural and charge requirements for activity. Biochim Biophys Acta 1462:29–54
73. Yasin B, Wang W, Pang M, Cheshenko N, Hong T, Waring AJ, Herold BC, Wagar EA, Lehrer RI (2004) Theta defensins protect cells from infection by herpes simplex virus by inhibiting viral adhesion and entry. J Virol 78:5147–5156
74. Song W, Shi Y, Xiao M, Lu H, Qu T, Li P, Wu G, Tian Y (2009) In vitro bactericidal activity of

recombinant human beta-defensin-3 against pathogenic bacterial strains in human tooth root canal. Int J Antimicrob Agents 33:237–243
75. Pachón-Ibáñez ME, Smani Y, Pachón J, Sánchez-Céspedes J (2017) Perspectives for clinical use of engineered human host defense antimicrobial peptides. FEMS Microbiol Rev 41:323–342
76. Sinha S, Cheshenko N, Lehrer RI, Herold BC (2003) NP-1, a rabbit alpha-defensin, prevents the entry and intercellular spread of herpes simplex virus type 2. Antimicrob Agents Chemother 47:494–500
77. Contreras X, Mzoughi O, Gaston F, Peterlin MB, Bahraoui E (2012) Protein kinase C-delta regulates HIV-1 replication at an early post-entry step in macrophages. Retrovirology 9:37
78. El-Kassas M, Awad A (2022) Metabolic aspects of hepatitis C virus. World J Gastroenterol 28:2429–2436
79. Gerhards C, Teufel A, Gerigk M, French M, Antoni C, Ebert M, Neumaier M, Evliyaoglu O (2024) Potential role of vitamin D in immune response in patients with viral hepatitis. Nutrition 124:112447
80. Huang J-F, Ko Y-M, Huang C-F, Yeh M-L, Dai C-Y, Hsieh M-H, Huang C-I, Yang H-L, Wang S-C, Lin Z-Y, Chen S-C, Yu M-L, Chuang W-L (2017) 25-Hydroxy vitamin D suppresses hepatitis C virus replication and contributes to rapid virological response of treatment efficacy. Hepatol Res 47: 1383–1389
81. Petta S, Cammà C, Scazzone C, Tripodo C, Di Marco V, Bono A, Cabibi D, Licata G, Porcasi R, Marchesini G, Craxí A (2010) Low vitamin D serum level is related to severe fibrosis and low responsiveness to interferon-based therapy in genotype 1 chronic hepatitis C. Hepatology 51:1158–1167
82. Villar LM, Del Campo JA, Ranchal I, Lampe E, Romero-Gomez M (2013) Association between vitamin D and hepatitis C virus infection: a meta-analysis. World J Gastroenterol 19:5917–5924
83. Kim HB, Myung SK, Lee YJ, Park BJ (2018) Efficacy of vitamin D supplementation in combination with conventional antiviral therapy in patients with chronic hepatitis C infection: a meta-analysis of randomised controlled trials. J Hum Nutr Diet 31: 168–177
84. Telcian AG, Zdrenghea MT, Edwards MR, Laza-Stanca V, Mallia P, Johnston SL, Stanciu LA (2017) Vitamin D increases the antiviral activity of bronchial epithelial cells in vitro. Antivir Res 137: 93–101
85. Jensen ME, Mailhot G, Alos N, Rousseau E, White JH, Khamessan A, Ducharme FM (2016) Vitamin D intervention in preschoolers with viral-induced asthma (DIVA): a pilot randomised controlled trial. Trials 17:353–353
86. Majak P, Olszowiec-Chlebna M, Smejda K, Stelmach I (2011) Vitamin D supplementation in children may prevent asthma exacerbation triggered by acute respiratory infection. J Allergy Clin Immunol 127:1294–1296
87. Yadav M, Mittal K (2014) Effect of vitamin D supplementation on moderate to severe bronchial asthma. Indian J Pediatr 81:650–654
88. Aregbesola A, Voutilainen S, Nurmi T, Virtanen JK, Ronkainen K, Tuomainen TP (2013) Serum 25-hydroxyvitamin D3 and the risk of pneumonia in an ageing general population. J Epidemiol Community Health 67:533–536
89. Mamani M, Muceli N, Ghasemi Basir HR, Vasheghani M, Poorolajal J (2017) Association between serum concentration of 25-hydroxyvitamin D and community-acquired pneumonia: a case-control study. Int J Gen Med 10:423–429
90. Brance ML, Miljevic JN, Tizziani R, Taberna ME, Grossi GP, Toni P, Valentini E, Trepat A, Zaccardi J, Moro J, Finuci Curi B, Tamagnone N, Ramirez M, Severini J, Chiarotti P, Consiglio F, Piñeski R, Ghelfi A, Kilstein J, Street E, Moretti D, Oliveto V, Mariño M, Manera J, Brun LR (2018) Serum 25-hydroxyvitamin D levels in hospitalized adults with community-acquired pneumonia. Clin Respir J 12:2220–2227
91. Wang TT, Nestel FP, Bourdeau V, Nagai Y, Wang Q, Liao J, Tavera-Mendoza L, Lin R, Hanrahan JW, Mader S, White JH (2004) Cutting edge: 1,25-dihydroxyvitamin D3 is a direct inducer of antimicrobial peptide gene expression. J Immunol 173:2909–2912
92. Lin R, White JH (2004) The pleiotropic actions of vitamin D. BioEssays 26:21–28
93. Jones BG, Oshansky CM, Bajracharya R, Tang L, Sun Y, Wong SS, Webby R, Thomas PG, Hurwitz JL (2016) Retinol binding protein and vitamin D associations with serum antibody isotypes, serum influenza virus-specific neutralizing activities and airway cytokine profiles. Clin Exp Immunol 183:239–247
94. Bals R, Wang X, Zasloff M, Wilson JM (1998) The peptide antibiotic LL-37/hCAP-18 is expressed in epithelia of the human lung where it has broad antimicrobial activity at the airway surface. Proc Natl Acad Sci USA 95:9541–9546
95. Mookherjee N, Brown KL, Bowdish DM, Doria S, Falsafi R, Hokamp K, Roche FM, Mu R, Doho GH, Pistolic J, Powers JP, Bryan J, Brinkman FS, Hancock RE (2006) Modulation of the TLR-mediated inflammatory response by the endogenous human host defense peptide LL-37. J Immunol 176:2455–2464
96. Abhishek BJ, Swamy AHV, Nyamagoud SB, George A, Namratha D (2024) Respiratory syncytial virus: an overview of clinical manifestations and management in the Indian pediatric population. Monaldi Arch Chest Dis. https://doi.org/10.4081/monaldi.2024.2888
97. Verma P, Chandelia S, Bhagat S, Selvakumar R, Kumari S, Mahich S, Patra B (2025) Severe

pneumonia and pediatric acute respiratory distress syndrome caused by HCoV NL 63 and respiratory syncytial virus coinfection with favorable outcome. Pediatr Infect Dis J. https://doi.org/10.1097/INF.0000000000004846
98. Zdrenghea MT, Makrinioti H, Bagacean C, Bush A, Johnston SL, Stanciu LA (2017) Vitamin D modulation of innate immune responses to respiratory viral infections. Rev Med Virol 27:e1909
99. Hansdottir S, Monick MM, Lovan N, Powers L, Gerke A, Hunninghake GW (2010) Vitamin D decreases respiratory syncytial virus induction of NF-kappaB-linked chemokines and cytokines in airway epithelium while maintaining the antiviral state. J Immunol 184:965–974
100. Martineau AR, Jolliffe DA, Hooper RL, Greenberg L, Aloia JF, Bergman P, Dubnov-Raz G, Esposito S, Ganmaa D, Ginde AA, Goodall EC, Grant CC, Griffiths CJ, Janssens W, Laaksi I, Manaseki-Holland S, Mauger D, Murdoch DR, Neale R, Rees JR, Simpson S Jr, Stelmach I, Kumar GT, Urashima M, Camargo CA Jr (2017) Vitamin D supplementation to prevent acute respiratory tract infections: systematic review and meta-analysis of individual participant data. BMJ 356:i6583
101. Rech MA, Fleming JN, Moore CL (2014) 25-hydroxyvitamin D deficiency and opportunistic viral infections after kidney transplant. Exp Clin Transplant 12:95–100
102. Gruber-Bzura BM (2018) Vitamin D and influenza—prevention or therapy? Int J Mol Sci 19:2419
103. Jia H, Sheng F, Yan Y, Liu X, Zeng B (2024) Vitamin D supplementation for prevention of acute respiratory infections in older adults: a systematic review and meta-analysis. PLoS One 19:e0303495
104. Cannell JJ, Vieth R, Umhau JC, Holick MF, Grant WB, Madronich S, Garland CF, Giovannucci E (2006) Epidemic influenza and vitamin D. Epidemiol Infect 134:1129–1140
105. Sharifi A, Vahedi H, Nedjat S, Rafiei H, Hosseinzadeh-Attar MJ (2019) Effect of single-dose injection of vitamin D on immune cytokines in ulcerative colitis patients: a randomized placebo-controlled trial. APMIS 127:681–687
106. Lang PO, Samaras D (2012) Aging adults and seasonal influenza: does the vitamin D status (h)arm the body? J Aging Res 2012:806198
107. Khare D, Godbole NM, Pawar SD, Mohan V, Pandey G, Gupta S, Kumar D, Dhole TN, Godbole MM (2013) Calcitriol [1, 25 [OH] 2 D3] pre-and post-treatment suppresses inflammatory response to influenza A (H1N1) infection in human lung A549 epithelial cells. Eur J Nutr 52:1405–1415
108. Helming L, Böse J, Ehrchen J, Schiebe S, Frahm T, Geffers R, Probst-Kepper M, Balling R, Lengeling A (2005) 1alpha,25-Dihydroxyvitamin D3 is a potent suppressor of interferon gamma-mediated macrophage activation. Blood 106:4351–4358
109. Moan JE, Dahlback A, Ma L, Juzeniene A (2009) Influenza, solar radiation and vitamin D. Dermatoendocrinol 1:308–310
110. Laaksi I, Ruohola J-P, Mattila V, Auvinen A, Ylikomi T, Pihlajamäki H (2010) Vitamin D supplementation for the prevention of acute respiratory tract infection: a randomized, double-blinded trial among young Finnish men. J Infect Dis 202:809–814
111. Ginde AA, Mansbach JM, Camargo CA (2009) Association between serum 25-hydroxyvitamin D level and upper respiratory tract infection in the Third National Health and Nutrition Examination Survey. Arch Intern Med 169:384–390
112. Berry DJ, Hesketh K, Power C, Hyppönen E (2011) Vitamin D status has a linear association with seasonal infections and lung function in British adults. Br J Nutr 106:1433–1440
113. Nanri A, Nakamoto K, Sakamoto N, Imai T, Akter S, Nonaka D, Mizoue T (2017) Association of serum 25-hydroxyvitamin D with influenza in case-control study nested in a cohort of Japanese employees. Clin Nutr 36:1288–1293
114. White JH (2022) Emerging roles of vitamin D-induced antimicrobial peptides in antiviral innate immunity. Nutrients 14:284
115. Ahsan N, Imran M, Mohammed Y, Al Anouti F, Khan MI, Banerjee T, Adnan M, Ashfaq F, Kieliszek M, Ashraf SA, Haq A (2023) Mechanistic insight into the role of vitamin D and zinc in modulating immunity against COVID-19: a view from an immunological standpoint. Biol Trace Elem Res 201:5546–5560
116. Beard JA, Bearden A, Striker R (2011) Vitamin D and the anti-viral state. J Clin Virol 50:194–200
117. Ismailova A, White JH (2022) Vitamin D, infections and immunity. Rev Endocr Metab Disord 23:265–277
118. Sousa FH, Casanova V, Findlay F, Stevens C, Svoboda P, Pohl J, Proudfoot L, Barlow PG (2017) Cathelicidins display conserved direct antiviral activity towards rhinovirus. Peptides 95:76–83
119. Honardoost M, Maryam G, Khamseh ME (2023) Role of vitamin D in pathogenesis and severity of COVID-19 infection. Arch Physiol Biochem 129:26–32
120. Schögler A, Muster RJ, Kieninger E, Casaulta C, Tapparel C, Jung A, Moeller A, Geiser T, Regamey N, Alves MP (2016) Vitamin D represses rhinovirus replication in cystic fibrosis cells by inducing LL-37. Eur Respir J 47:520–530
121. Grant WB, Lahore H, McDonnell SL, Baggerly CA, French CB, Aliano JL, Bhattoa HP (2020) Evidence that vitamin D supplementation could reduce risk of influenza and COVID-19 infections and deaths. Nutrients 12:988

122. Niv Y (2020) Defensin 5 for prevention of SARS-CoV-2 invasion and Covid-19 disease. Med Hypotheses 143:110244
123. Wang C, Wang S, Li D, Wei DQ, Zhao J, Wang J (2020) Human intestinal defensin 5 inhibits SARS-CoV-2 invasion by cloaking ACE2. Gastroenterology 159:1145–1147.e1144
124. Wang C, Wang S, Li D, Chen P, Han S, Zhao G, Chen Y, Zhao J, Xiong J, Qiu J, Wei DQ, Zhao J, Wang J (2021) Human cathelicidin inhibits SARS-CoV-2 infection: killing two birds with one stone. ACS Infect Dis 7:1545–1554
125. Zhang L, Ghosh SK, Basavarajappa SC, Muller-Greven J, Penfield J, Brewer A, Ramakrishnan P, Buck M, Weinberg A (2021) Molecular dynamics simulations and functional studies reveal that hBD-2 binds SARS-CoV-2 spike RBD and blocks viral entry into ACE2 expressing cells. bioRxiv 2021: 425621. https://doi.org/10.1101/2021.01.07.425621
126. Greiller CL, Martineau AR (2015) Modulation of the immune response to respiratory viruses by vitamin D. Nutrients 7:4240–4270. https://doi.org/10.3390/nu7064240

MIX
Papier aus verantwortungsvollen Quellen
Paper from responsible sources
FSC® C105338

If you have any concerns about our products,
you can contact us on
ProductSafety@springernature.com

In case Publisher is established outside the EU,
the EU authorized representative is:
Springer Nature Customer Service Center GmbH
Europaplatz 3, 69115 Heidelberg, Germany

Printed by Libri Plureos GmbH
in Hamburg, Germany